The Carus Mathematical Monographs

Number Twenty-Three

COMPLEX ANALYSIS: THE GEOMETRIC VIEWPOINT

Steven G. Krantz
Washington University in St. Louis

Published and Distributed by
THE MATHEMATICAL ASSOCIATION OF AMERICA

Complex Analysis: The Geometric Viewpoint

Steven G. Krantz

THE
CARUS MATHEMATICAL MONOGRAPHS

Published by
THE MATHEMATICAL ASSOCIATION OF AMERICA

The following Monographs have been published:

1. *Calculus of Variations,* by G. A. Bliss
2. *Analytic Functions of a Complex Variable,* by D. R. Curtiss
3. *Mathematical Statistics,* by H. L. Rietz
4. *Projective Geometry,* by J. W. Young
5. *A History of Mathematics in America before 1900,* by D. E. Smith and Jekuthiel Ginsburg (out of print)
6. *Fourier Series and Orthogonal Polynomials,* by Dunham Jackson
7. *Vectors and Matrices,* by C. C. MacDuffee
8. *Rings and Ideals,* by N. H. McCoy
9. *The Theory of Algebraic Numbers,* second edition, by Harry Pollard and Harold G. Diamond
10. *The Arithmetic Theory of Quadratic Forms,* by B. W. Jones
11. *Irrational Numbers,* by Ivan Niven
12. *Statistical Independence in Probability, Analysis and Number Theory,* by Mark Kac
13. *A Primer of Real Functions,* third edition, by Ralph P. Boas, Jr.
14. *Combinatorial Mathematics,* by Herbert J. Ryser
15. *Noncommutative Rings,* by I. N. Herstein
16. *Dedekind Sums,* by Hans Rademacher and Emil Grosswald
17. *The Schwarz Function and its Applications,* by Philip J. Davis
18. *Celestial Mechanics,* by Harry Pollard
19. *Field Theory and its Classical Problems,* by Charles Robert Hadlock
20. *The Generalized Riemann Integral,* by Robert M. McLeod
21. *From Error-Correcting Codes through Sphere Packings to Simple Groups,* by Thomas M. Thompson
22. *Random Walks and Electric Networks,* by Peter G. Doyle and J. Laurie Snell

Library of Congress Catalog Card Number 90-070795

Complete Set ISBN 0-88385-000-1
Vol. 23 ISBN 0-88385-026-5

Printed in the United States of America

Current Printing (last digit):
10 9 8 7 6 5 4 3 2 1

To my parents.

ACKNOWLEDGEMENTS

This book owes its existence to many people. I thank Don Albers, Robert E. Greene, and Paul Halmos for convincing me to write it. I am grateful to Marco Abate, Harold Boas, Ralph Boas, David Drasin, Paul Halmos, Daowei Ma, David Minda, Marco Peloso, John Stapel, and Jim Walker for reading various versions of the manuscript and making valuable comments and suggestions. The Carus monograph committee of the MAA helped me to find the right level and focus for the book.

Ralph Boas, the chairman of the Carus monograph committee, played a special role in the development of this book. In addition to shepherding the project along, he provided necessary prodding and cajoling at crucial stages to keep the project on track. For this I express my sincere gratitude.

I thank Jim Milgram for creating TECPRINT, the world's first technical word processor. This has been a valuable tool to me for many years. I am grateful to Micki Wilderspin for performing the tedious task of converting the manuscript from TECPRINT into TEX. Beverly Joy Ruedi did a splendid job of making my project into a finished book.

A special vote of thanks goes to Robert Burckel for reading most of the manuscript inch by inch and correcting both my mathematics and my writing. His contributions have improved the quality of the book decisively. Responsibility for all remaining errors of course resides entirely with me.

—S.G.K.

PREFACE

The modern geometric point of view in complex function theory began with Ahlfors's classic paper [AHL] . In that work it was demonstrated that the Schwarz lemma can be viewed as an inequality of certain differential geometric quantities on the disc (we will later learn that they are curvatures). This point of view—that substantive analytic facts can be interpreted in the language of Riemannian geometry—has developed considerably in the last fifty years. It provides new proofs of many classical results in complex analysis, and has led to new insights as well.

In this monograph we intend to introduce the reader with a standard one semester background in complex analysis to the geometric method. All geometric ideas will be developed from first principles, and only to the extent needed here. No background in geometry is assumed or required.

Chapter 0 gives a bird's eye view of classical function theory of one complex variable. We pay special attention to topics which are developed later in the book from a more advanced point of view. In this chapter we also sketch proofs of the main results, with the hope that the reader can thereby get a feeling for classical methodology before embarking on a study of the geometric method.

Chapter 1 begins a systematic treatment of the techniques of Riemannian geometry, specially tailored to the setting of one complex variable. In order that the principal ideas may be brought out most clearly, we shall concentrate on only a few themes: the Schwarz lemma, the Riemann mapping theorem, normal families, and Picard's theorems. For many readers this will be a first contact with the latter two results. The geometric method provides a particularly cogent explanation of these theorems, and can be contrasted with the more classical proofs which are discussed in Chapter 0. We shall also touch on Fatou-Julia theory, a topic which is rather technical from the analytic standpoint but completely natural from the point of view of geometry.

In Chapter 3 we introduce the Carathéodory and Kobayashi metrics, a device which is virtually unknown in the world of one complex variable. This decision allows us to introduce invariant metrics on arbitrary planar domains without resort to the uniformization theorem. We are then able to give a "differential geometric" interpretation of the Riemann mapping theorem.

The last chapter gives a brief glimpse of several complex variables. Some of the themes which were developed earlier in the book are carried over to two dimensions. Biholomorphic mappings are discussed, and the inequivalence of the ball and bidisc is proved using a geometric argument.

The language of differential geometry is not generally encountered in a first course in complex analysis. It is hoped that this volume will be used as a supplement to such a course, and that it may lead to greater familiarity with the fruitful methodology of geometry.

TABLE OF CONTENTS

CHAPTER 0
PRINCIPAL IDEAS OF CLASSICAL FUNCTION THEORY

1. A Glimpse of Complex Analysis

The purpose of this book is to explain how various aspects of complex analysis can be understood both naturally and elegantly from the point of view of metric geometry. Thus, in order to set the stage for our work, we begin with a review of some of the principal ideas in complex analysis. A good companion volume for this introductory material is [BOAS]. See also [KR3].

Central to the subject are the Cauchy integral theorem and the Cauchy integral formula. From these follow the Cauchy estimates, Liouville's theorem, the maximum principle, Schwarz's lemma, the argument principle, Montel's theorem, and most of the other powerful and elegant results which are basic to the subject. We will discuss these results in essay form. The proofs that we provide are more conceptual than rigorous: *the aim is to depict a flow of ideas rather than absolute mathematical precision.*

If $P \in \mathbf{C}$ and $r > 0$, then we use the standard notation

$$\begin{aligned} D(P,r) &= \left(z \in \mathbf{C} : |z-P| < r\right), \\ \bar{D}(P,r) &= \left(z \in \mathbf{C} : |z-P| \le r\right), \\ \partial D(P,r) &= \left(z \in \mathbf{C} : |z-P| = r\right). \end{aligned}$$

A connected open set $U \subseteq \mathbf{C}$ is called a *domain.*

Complex analysis consists of the study of holomorphic functions. Let F be a complex-valued continuously differentiable function (in the sense of multivariable calculus) on a domain U in the complex plane. We write $F = u + iv$ to distinguish the real and imaginary parts of F. Then F is said to be *holomorphic*, or *analytic*, if it satisfies the Cauchy–Riemann equations:

$$\frac{\partial u}{\partial x} = \frac{\partial v}{\partial y} \qquad \text{and} \qquad \frac{\partial u}{\partial y} = -\frac{\partial v}{\partial x}.$$

This definition is equivalent to other familiar definitions, such as that in terms of the complex derivative, which we now discuss. If F is a function on a domain U in the complex plane and if $P \in U$, then F is said to possess a complex derivative, or to be complex differentiable, at P if

$$F'(P) = \frac{\partial F}{\partial z}(P) = \lim_{z \to P} \frac{F(z) - F(P)}{z - P}$$

exists. The function F is holomorphic on U if F possesses the complex derivative at each point of U.

The complex derivative definition of "holomorphic" is of great historical interest. Much effort was expended in the early days of the subject in proving that a function which is complex differentiable at each point of a domain U is in fact continuously differentiable (in the usual sense of multivariable calculus), and from that point it is routine to check that the function satisfies the Cauchy–Riemann equations. The converse implication is a straightforward exercise. So, in the end, either definition is correct. From our perspective the Cauchy–Riemann equations provide the most useful point of view. This assertion will become more transparent as we develop the notion of complex integration.

Definition 1.

A C^1, or *continuously differentiable,* curve in a domain $U \subseteq \mathbf{C}$ is a function $\gamma : [a,b] \to U$ from an interval in the real line into U such that γ' exists at each point of $[a,b]$ (in the one-sided sense at the endpoints) and is continuous on $[a,b]$. When there is no danger of confusion, we sometimes use the symbol γ to denote the set of points $\{\gamma(t) : t \in [a,b]\}$ rather than to denote the function from $[a,b]$ to U.

A *piecewise continuously differentiable curve* is a single continuous curve which can be written as a finite union of continuously differentiable curves. A curve is called *closed* if $\gamma(a) = \gamma(b)$. It is called *simple closed* if it is closed and not self-intersecting: $\gamma(s) = \gamma(t)$ and $s \neq t$ imply either that $s = a$ and $t = b$ or that $s = b$ and $t = a$.

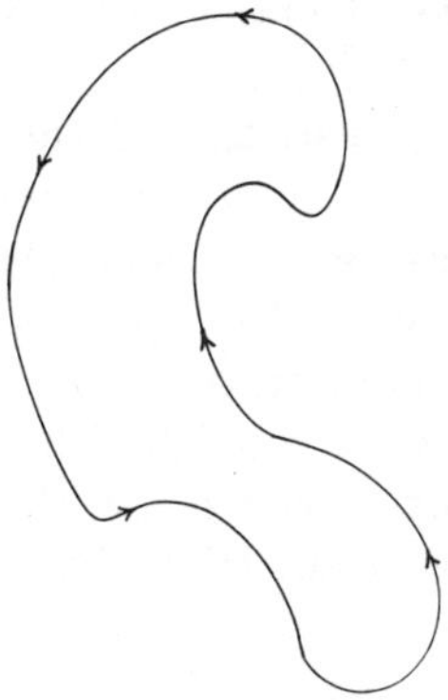

FIGURE 1

A simple closed curve γ is said to be *positively oriented* if the region interior to the curve is to the left of the curve while it is being traversed from $t = a$ to $t = b$. See Figure 1. Otherwise it is called *negatively oriented.* If F is a continuous function on our open set U, then we define its *complex line integral* over a continuously differentiable curve γ in U to be the quantity

$$\oint_\gamma F(z)\,dz = \int_a^b F(\gamma(t)) \cdot \gamma'(t)\,dt.$$

Notice that, in analogy with the study of directed curves in Stokes's formula, the derivative of the curve is incorporated into the integral. In case γ is only piecewise continuously differentiable, we define

$$\oint_\gamma F(z)\,dz$$

by integrating along each of the continuously differentiable pieces and adding.

Now we may formulate the Cauchy integral theorem. A rigorous treatment of this result requires a discussion of deformation of curves. However, since this is only a review, we may be a bit imprecise. Let γ be a curve in a domain U and suppose that γ (more precisely, the *image* of γ) can be continuously deformed to a point within U. We shall call such a curve "topologically trivial (with respect to U)." In Figure 2, γ_2 is topologically trivial but γ_1 is not. Now we have:

□ **Theorem 2.** *Let F be a holomorphic function on a domain U and let γ be a topologically trivial, piecewise continuously differentiable, closed curve in U. Then*

$$\oint_\gamma F(z)\,dz = 0.$$

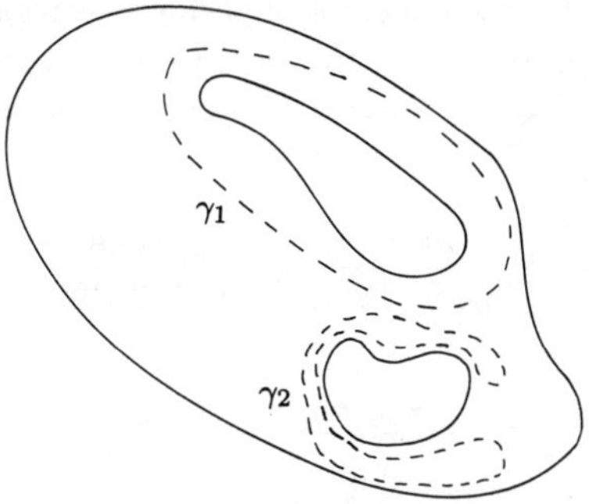

FIGURE 2

This theorem may be proved, using the Cauchy-Riemann equations, as a direct application of Stokes's theorem. It will tell us, in effect, that a holomorphic function is strongly influenced on an open set by its behavior at the boundary of that set.

Now fix a point P in U and let γ be a positively oriented, topologically trivial, *simple closed curve* in U with P in its interior. Let F be holomorphic on U. Now, by suitable limiting arguments, we may apply the Cauchy integral theorem to the function

$$G(z) \equiv \begin{cases} \dfrac{F(z) - F(P)}{z - P} & \text{if } z \neq P, \\ F'(P) & \text{if } z = P. \end{cases}$$

After some calculations, the result is that

$$F(z) = \frac{1}{2\pi i} \oint_\gamma \frac{F(\zeta)}{\zeta - z}\, d\zeta .$$

This is the *Cauchy integral formula.* It vindicates our anticipation that a holomorphic function is completely determined in the interior of γ by its behavior on the boundary curve γ itself. From this there quickly flows a wealth of information.

□ **Theorem 3.** *Let F be holomorphic on a domain U and let $P \in U$. Assume that the closed disc*

$$\bar{D}(P,r) \equiv \{z : |z - P| \leq r\}$$

is contained in U. Then F may be written on $\bar{D}(P,r)$ as a convergent power series:

$$F(z) = \sum_{j=0}^{\infty} a_j (z - P)^j .$$

The convergence is absolute and uniform on the disc $\bar{D}(P,r)$.

Thus we see that, in a natural sense, holomorphic functions are generalizations of complex polynomials. The power series expansion is, in general, only local. But for many purposes this is sufficient.

PROOF OF THEOREM 3.
Observe that, for $|z - P| < r$ and $|\zeta - P| = r$, we may write

$$\frac{1}{\zeta - z} = \frac{1}{\zeta - P} \cdot \frac{1}{1 - \frac{z-P}{\zeta-P}} .$$

Since $|z - P| < r = |z - \zeta|$, we have that

$$\left| \frac{z - P}{\zeta - P} \right| < 1.$$

Thus

$$\frac{1}{\zeta - z} = \frac{1}{\zeta - P} \cdot \sum_{j=0}^{\infty} \left(\frac{z - P}{\zeta - P} \right)^j .$$

Substituting this power series expansion for the Cauchy kernel into the Cauchy integral formula on $\bar{D}(P,r)$ gives the desired power series expansion for the holomorphic function F. □

As an added bonus, the proof gives us a formula for the series coefficients a_j:

$$a_j = \frac{1}{2\pi i} \int_\gamma \frac{F(\zeta)}{(\zeta - P)^{j+1}} \, d\zeta .$$

Just as in the theory of Taylor series, it turns out that the coefficients a_j must also be given by

$$a_j = \frac{1}{j!} \left(\frac{\partial^j F}{\partial z^j} \right) (P).$$

We conclude that

$$\left(\frac{\partial^j F}{\partial z^j} \right) (P) = \frac{j!}{2\pi i} \oint_\gamma \frac{F(\zeta)}{(\zeta - P)^{j+1}} \, d\zeta . \qquad (*)$$

□ **Corollary.** *Let $\tilde{F}$ be a holomorphic function on a punctured disc $\tilde{D}(P,r) \equiv D(P,r) \setminus \{P\}$. If $\tilde{F}$ is bounded then $\tilde{F}$ continues analytically to the entire disc $D(P,r)$. That is, there is a holomorphic function F on $D(P,r)$ such that $F\big|_{\tilde{D}(P,r)} = \tilde{F}$.*

SKETCH OF PROOF.
Assume without loss of generality that $P = 0$. Consider the function $G(z)$ which is defined to equal $z^2 \cdot \tilde{F}$ on $\tilde{D}(P,r)$ and to equal 0 at $P = 0$. Then G is continuously differentiable on $D(P,r)$ and satisfies the Cauchy-Riemann equations.

The leading term of the power series expansion of G about 0 is of the form $a_2 z^2$. Thus the holomorphic function G may be divided by z^2 to define a holomorphic function F on $D(P,r)$ that agrees with $\tilde{F}$ on $\tilde{D}(P,r)$. □

It is a standard fact from the theory of power series that the zeroes of a function given by a power series expansion cannot accumulate in the interior of the domain of that function. Thus we have:

□ **Theorem 4.** *If F is holomorphic on a domain U, then $\{z \in U : F(z) = 0\}$ has no accumulation point in U.*

This theorem once again bears out the dictum that holomorphic functions are much like polynomials: The zero set of a polynomial $a_0 + a_1 z + a_2 z^2 + \cdots + a_n z^n$ is discrete, indeed it is finite.

The Cauchy estimates on the derivatives of a holomorphic function follow from direct estimation of the formula (*):

□ **Theorem 5.** *Let F be a holomorphic function on a domain U that contains the closed disc $\bar{D}(P,R)$. Then the derivatives of F satisfy the estimates*

$$\left| \left(\frac{\partial^j}{\partial z^j} \right) F(P) \right| \leq \frac{j! \cdot M}{R^j},$$

where M is the supremum of $|F|$ on the disc $\bar{D}(P,r)$.

Notice that this result tells us that if F is bounded on a large disc, then its derivatives are relatively small; this assertion is exploited in the next result.

□ **Theorem 6** (Liouville). *Let F be a holomorphic function on the complex plane (an entire function) which is also bounded. Then F must be a constant.*

PROOF.
Assume without loss of generality that $|F|$ is bounded by 1. Fix a point P in the plane. Applying the Cauchy estimates to F on the disc $D(P,R)$ yields that

$$|F'(P)| \leq \frac{1}{R^1}.$$

Letting R tend to $+\infty$ yields that $F' \equiv 0$. It is now a simple calculus exercise to see that F must be constant. □

One of the most dramatic applications of Liouville's theorem is in the proof of the fundamental theorem of algebra. That is our next task:

□ **Theorem 7.** *Let $p(z) = a_0 + a_1 z + a_2 z^2 + \cdots + a_n z^n$ be a polynomial of degree at least one. Then there is a point z at which p vanishes.*

PROOF.
Suppose not. Then $F(z) = 1/p(z)$ is an entire function. Since a nonconstant polynomial blows up at infinity, F must be bounded. By Liouville's theorem, F is a constant. Hence p is constant, and therefore has degree zero. This contradiction completes the proof. □

Let k be the degree of the polynomial p. Notice that if the polynomial p vanishes at the point r_1, then p is divisible by $(z - r_1) : p(z) = (z - r_1) \cdot p_1(z)$ for a polynomial p_1 of degree $k - 1$. If $k - 1 \geq 1$, then we may apply the last result to p_1. Continuing in this fashion, we obtain that p may be expressed as a product of linear factors:

$$p(z) = (z - r_1) \cdot (z - r_2) \cdots (z - r_k).$$

We conclude this brief overview of elementary complex analysis by recalling the argument principle and Hurwitz's theorem.

□ **Theorem 8** (The Argument Principle). *Let F be holomorphic on a domain U and let γ be a topologically trivial, positively oriented, simple closed curve in U. Assume that F does not vanish on γ. We can be sure, by Theorem 4, that there are at most finitely many, say k, zeroes of F inside γ (counting multiplicity). Then it holds that*

$$k = \frac{1}{2\pi i} \oint_\gamma \frac{F'(\zeta)}{F(\zeta)} \, d\zeta .$$

SKETCH OF PROOF.
By an easy reduction, it is enough to prove the result when $k = 1$. A second reduction reduces to the case when γ is a positively oriented circle. After a change of coordinates, let us suppose that F has a simple zero at the point $P = 0$ inside γ. By writing out the power series expansions for F and for F', we find that

$$\frac{F'(\zeta)}{F(\zeta)} = \frac{1}{\zeta} + h(\zeta),$$

where h is holomorphic near 0. Of course, h integrates to 0, by the Cauchy integral theorem. And it is easily calculated that

$$\frac{1}{2\pi i}\oint_\gamma \frac{1}{\zeta}\, d\zeta = 1,$$

completing the proof. □

Let U be a domain and $\{F_j\}$ a sequence of holomorphic functions on U which converges, uniformly on compact sets, to a limit function F. It is an easy consequence of the Cauchy integral formula that the limit function is also holomorphic. Let us now use the argument principle to see how the zeroes of F are related to the zeroes of the F_j's.

□ **Theorem 9** (Hurwitz's Theorem). *With $\{F_j\}$ and F as above, if the F_j's are all zero free, then either F is zero free or F is identically zero.*

PROOF.
Assume that F is not identically 0. Then the zero set of F is discrete, so we can find a simple, closed, topologically trivial γ in U which misses the zero set completely. Therefore,

$$\frac{1}{2\pi i}\oint_\gamma \frac{F_j'(\zeta)}{F_j(\zeta)}\, d\zeta \longrightarrow \frac{1}{2\pi i}\oint_\gamma \frac{F'(\zeta)}{F(\zeta)}\, d\zeta$$

as $j \to \infty$. But the expression on the left is zero for each j. Hence so is the expression on the right. So F has no zeroes inside the curve γ. Since γ was chosen to be an "arbitrary" curve that misses the zeroes of F, we find that F is zero free. □

2. The Maximum Principle, the Schwarz Lemma, and Applications

We begin with a brief treatment of the maximum principle. Let F be holomorphic on an open set U which contains the closed disc $\bar{D}(P,r)$. Then the Cauchy integral formula says that

$$F(P) = \frac{1}{2\pi i} \oint_{\partial D(P,r)} \frac{f(\zeta)}{\zeta - P}\, d\zeta.$$

Parametrizing the boundary of the disc by $\gamma(t) = P + \mathrm{r}e^{it}$ and writing out the definition of the line integral gives

$$F(P) = \frac{1}{2\pi} \int_0^{2\pi} F(P + \mathrm{r}e^{it})\, dt.$$

This is the mean value property for a holomorphic function. Now we have:

□ **Theorem 1.** *Let F be holomorphic on a connected open set U. If there is a point $P \in U$ such that*

$$|F(P)| \geq |F(z)|$$

for all $z \in U$, then F is a constant function.

SKETCH OF PROOF.
Multiplying F by a constant, we may assume that $M \equiv F(P)$ is real and nonnegative. Let $S = \{z \in U : F(z) = F(P)\}$. Observe that S is nonempty since $P \in S$. Moreover, since F is a continuous function, S is trivially closed.

To see that S is open, let $w \in S$ and suppose that $D(w,r) \subseteq U$. Now, for $0 < r' < r$, it holds that

$$M = F(w) = \left| \frac{1}{2\pi} \int_0^{2\pi} F(P + r'e^{it})\, dt \right|$$
$$\leq \frac{1}{2\pi} \int_0^{2\pi} \left| F(P + r'e^{it}) \right| dt \leq M .$$

Since the expressions on the extreme left and right are equal, all the inequalities must be equalities. Thus, $F(P + r'e^{it}) = \left| F(P + r'e^{it}) \right| = M$ for all t and all $0 < r' < r$. Therefore,

$$\left\{ r'e^{it} : 0 \leq t \leq 2\pi ,\, 0 < r' < r \right\} \subseteq S .$$

This shows that S is open.

Since S is nonempty, closed, and open, and since U is connected, it follows that $S = U$. That is what we are required to prove. □

For us, the main application of the maximum principle will be to the classical Schwarz lemma.

□ **Theorem 2** (The Schwarz Lemma). *If $f : D \to D$ is holomorphic and if $F(0) = 0$, then*

$$\left| F(z) \right| \leq |z| \qquad \textit{and} \qquad \left| F'(0) \right| \leq 1 .$$

If either $\left| F(z) \right| = |z|$ for some $z \neq 0$ or $\left| F'(0) \right| = 1$, then F is a rotation: $F(z) \equiv e^{i\tau} \cdot z$ for some $\tau \in \mathbf{R}$.

PROOF.
The function

$$G(z) = \begin{cases} \dfrac{F(z)}{z}, & z \neq 0 \\ F'(z), & z = 0 \end{cases}$$

is holomorphic on D. Applying the maximum principle to G on the disc $\{z : |z| \leq 1 - \varepsilon\}$ for $\varepsilon > 0$ gives

$$|G(z)| \leq (1 - \varepsilon)^{-1}$$

when $|z| \leq 1 - \varepsilon$. Letting $\varepsilon \to 0^+$ yields $|G(z)| \leq 1$ on D, which is equivalent to the desired conclusion. One similarly obtains that $|F'(0)| \leq 1$.

For uniqueness, notice that $|F(z)| = |z|$ for some $z \neq 0$ implies that $|G(z)| = 1$. The maximum principle then forces G to be a unimodular constant, hence F is a rotation. The uniqueness statement for the derivative is obtained similarly. □

As a simple application of the Schwarz lemma, we classify the conformal self-maps (that is, the one-to-one, onto holomorphic maps) of the unit disc to itself. We begin by claiming that if a is a complex number of modulus less than 1, then the Möbius transformation

$$\phi_a(\zeta) = \frac{\zeta - a}{1 - \bar{a}\zeta}$$

maps $D(0,1)$ to $D(0,1)$. To see this, notice that ϕ_a is certainly well defined and holomorphic in a neighborhood of $\bar{D}(0,1)$; moreover, $\phi_a(a) = 0$. If we prove that ϕ_a maps $\partial D(0,1)$ to $\partial D(0,1)$, then our claim will follow from the maximum principle. But for $|\zeta| = 1$ we have that

$$|\phi_a(\zeta)| = \left|\frac{\zeta - a}{1 - \bar{a}\zeta}\right| = \left|\frac{1}{\bar{\zeta}} \cdot \frac{\zeta - a}{1 - \bar{a}\zeta}\right|$$
$$= \left|\frac{\zeta - a}{\bar{\zeta} - \bar{a}}\right| = 1,$$

since $\zeta \cdot \bar{\zeta} = 1$. This proves that $\phi_a : D(0,1) \to D(0,1)$. It is straightforward to check that $(\phi_a)^{-1} = \phi_{-a}$, proving that ϕ_a is one-to-one and onto.

In addition to the functions ϕ_a we also have the rotations $\rho_\tau(\zeta) \equiv e^{i\tau} \cdot \zeta$, $\tau \in \mathbf{R}$, which map the disc in a one-to-one, onto fashion to the disc. The remarkable fact is that these two types of maps completely characterize all conformal self-maps of the disc.

□ **Theorem 3.** *Let F be a conformal self-map of the unit disc. Then there exists a complex number a of modulus less than 1 and a $\tau \in \mathbf{R}$ such that*

$$F(\zeta) = \phi_a \circ \rho_\tau(\zeta).$$

PROOF.
Let $F(0) = b$ and consider the function $G \equiv \phi_b \circ F$. Then G is a conformal self-map of the unit disc and $G(0) = 0$. The Schwarz lemma tells us that $|G'(0)| \leq 1$. But we may also apply the Schwarz lemma to G^{-1} to obtain that $|1/G'(0)| = |(G^{-1})'(0)| \leq 1$. We conclude that $|G'(0)| = 1$. The uniqueness part of the Schwarz lemma then says that $G(\zeta) = \rho_\tau(\zeta)$ for some τ. But this is equivalent to

$$F = \phi_{-b} \circ \rho_\tau .$$

Setting $a = -b$ completes the proof. □

Exercises.

Give a similar proof to show that every conformal self-map of the disc can be written in the form $\rho_\tau \circ \phi_a$.

Verify directly that the composition of two functions of the form $\phi_a \circ \rho_\tau$ (respectively, $\rho_\tau \circ \phi_a$) is also a function of that form.

Our next step is to generalize the Schwarz lemma by removing the requirement that $F(0) = 0$. We have:

□ **Theorem 4** (Schwarz, Pick). *If $F : D \to D$ is holomorphic, $F(z_1) = w_1$ and $F(z_2) = w_2$, then*

$$\left|\frac{w_1 - w_2}{1 - w_1 \bar{w}_2}\right| \le \left|\frac{z_1 - z_2}{1 - z_1 \bar{z}_2}\right|$$

and

$$|F'(z_1)| \le \frac{1 - |w_1|^2}{1 - |z_1|^2}.$$

If equality obtains in the first expression for some $z_1 \neq z_2$ or if equality obtains in the second expression, then F must be a conformal self-map of the disc.

PROOF.
Define

$$\phi(z) = \frac{z + z_1}{1 + \bar{z}_1 z}, \qquad \psi(z) = \frac{z - w_1}{1 - \bar{w}_1 z}.$$

Then $\psi \circ F \circ \phi$ satisfies the hypotheses of Schwarz's lemma. Therefore

$$\left|(\psi \circ F \circ \phi)(z)\right| \le |z|, \qquad \forall z \in D.$$

Setting $z = \phi^{-1}(z_2)$ now gives the first inequality. Also the Schwarz lemma says that

$$\left|(\psi \circ F \circ \phi)'(0)\right| \leq 1.$$

Using the chain rule to write this out gives the second inequality.

The case of equality is analyzed as in Theorem 2. □

3. Normal Families and the Riemann Mapping Theorem

One of the most important concepts in topology is compactness. Compactness for a set of points in Euclidean space is, thanks to the Heine-Borel theorem, easy to understand: a set is compact if and only if it is closed and bounded. In modern times, compactness for families of functions has proved to be a powerful tool. The notion of normal families, and in particular of Montel's theorem, is historically one of the first instances of this concept. In the present section we shall treat both Montel's theorem and its application to the Riemann mapping theorem.

In an effort to keep our exposition simple, we treat here a restrictive definition of normal family; later in the book normal families will be given a more thorough treatment from a completely different point of view.

Definition 1.

A family $\mathcal{F}$ of functions on a domain U will be called *normal* if every sequence in $\mathcal{F}$ has a subsequence which converges uniformly on compact subsets of U.

Now Montel's theorem says the following

□ **Theorem 2** (Montel). *Let U be a domain in the complex plane. Let $\mathcal{F}$ be a family of holomorphic functions on U. If there is a positive constant M such that*

$$|F(z)| \leq M$$

for all $z \in U$ and all $F \in \mathcal{F}$, then $\mathcal{F}$ is a normal family.

For the proof of this theorem, we shall need an important result from real analysis. We begin with a little terminology.

Definition 3.

Let $\mathcal{F}$ be a family of functions on a common domain $S \subseteq \mathbf{R}^n$. We say that the family is *equicontinuous* if for each $\varepsilon > 0$ there is a $\delta > 0$ such that whenever $z, w \in S$ satisfy $|z - w| < \delta$, then

$$|F(z) - F(w)| < \varepsilon$$

for all $F \in \mathcal{F}$.

Notice that the property of equicontinuity is stronger than uniform continuity: not only is the choice of δ independent of the points z and w in S, but it is independent of which F from the family $\mathcal{F}$ we are considering. Next we define a related notion.

Definition 4.

Let $\mathcal{F}$ be a family of functions on a common domain $S \subseteq \mathbf{R}^n$. We say that the family is *equibounded* if there is a number $M > 0$ such that, whenever $z \in S$ and $F \in \mathcal{F}$, then

$$|F(z)| \leq M.$$

We use the notions of equicontinuity and equiboundedness to formulate the following essential result.

□ **Proposition 5** (Ascoli/Arzela). *Let K be a compact set in $\mathbf{R}^N$. Let $\mathcal{F}$ be an equicontinuous family of functions on K which is also equibounded. Then $\mathcal{F}$ contains a sequence $\{f_j\}$ which converges uniformly on K.*

The Ascoli-Arzela theorem is too difficult and technical to prove here. We refer the interested reader to [RU1] for details. Note, however, the thrust of the theorem: if a family of rather nice objects is bounded, then it has a convergent subsequence. In other words, the theorem is a statement about compactness.

Now we apply the Ascoli/Arzela theorem to prove Montel's theorem:

PROOF OF THEOREM 2.

Let $\bar{D}(P,R) \subseteq U$. Since U is open, the complement ${}^cU$ of U is closed. Since $\bar{D}(P,R)$ and ${}^cU$ are disjoint, there is a positive distance between the two sets: that is, there exists a $c > 0$ such that if $z \in \bar{D}(P,R)$ and $u \in {}^cU$, then $|z - u| > c$. Thus, for any $z \in \bar{D}(P,R)$ and any $F \in \mathcal{F}$ we may apply the Cauchy estimates to F on the disc $\bar{D}(z,c)$.

We conclude that

$$|F'(z)| \le \frac{M \cdot 1!}{c^1} \equiv C.$$

It is now an easy exercise, using the mean value theorem, to see that

$$\big|F(z) - F(w)\big| \le C|z - w|$$

for any $z, w \in \bar{D}(P,R)$. This estimate shows that the family $\mathcal{F}$ is equicontinuous on $\bar{D}(P,R)$: given $\varepsilon > 0$, choose $\delta = \varepsilon/C$.

Finally, if K is any compact subset of U, then K can be covered by finitely many discs $\bar{D}(P,r)$ as above, and the equicontinuity on K follows from the triangle inequality.

We may now apply the Ascoli-Arzela theorem to $\mathcal{F}$ on K to find a uniformly convergent subsequence $\{F_{j_k}\}$ of any given sequence $\{F_j\} \subseteq \mathcal{F}$. A standard diagonalization argument then yields a subsequence $\{F_{j_{k_l}}\}$ that converges uniformly on *all* compact subsets of U. □

Notice that in real analysis there is no analogue for Montel's theorem. The family $\mathcal{F} = \{\sin jx\}_{j=1}^{\infty}$ is equibounded, as in the hypothesis of Montel's theorem, but there is no convergent subsequence. It is the property of being holomorphic that gives the required extra structure.

Now we turn our attention to the Riemann mapping theorem. Let U and V be open sets in the complex plane. We say that U and V are conformally equivalent if there is a one-to-one, onto holomorphic function (i.e., a *conformal map*)

$$\sigma : U \to V.$$

The motivation for this concept is clear: any holomorphic function F on V gives rise to a holomorphic function $F \circ$

σ on U and any holomorphic function G on U gives rise to a holomorphic function $G \circ \sigma^{-1}$ on V. Thus complex analysis on the two domains is, in effect, equivalent.

How can we tell when two domains are conformally equivalent? An obvious necessary condition is that they be topologically equivalent, for any conformal mapping is certainly a homeomorphism. Riemann's astonishing theorem is that if a domain U is topologically equivalent to the unit disc, and is not the entire plane, then it is conformally equivalent to the unit disc.

□ **Theorem 6** (the Riemann Mapping Theorem). *Let U be a proper subdomain of $\mathbf{C}$ which is homeomorphic to the unit disc. Then U is conformally equivalent to the open unit disc $D(0,1)$.*

The proof of this theorem is quite elaborate; we shall outline the principal steps. To keep matters as simple as possible, we shall assume that U is bounded.

>>> STEP 1

Fix a point $P \in U$. Let $\mathcal{F} = \{\sigma : U \to D(0,1)$ such that σ is one-to-one, holomorphic, and $\sigma(P) = 0\}$. We claim that $\mathcal{F}$ is nonempty. In fact, since U is bounded, there is an $R > 0$ such that $U \subseteq D(0,R)$. Then the function

$$\sigma : \zeta \mapsto \frac{1}{2R}(\zeta - P)$$

certainly maps P to 0, is holomorphic, and satisfies

$$|\sigma(\zeta)| < \frac{1}{2R}(R + R) = 1.$$

Obviously σ is one-to-one. So $\sigma \in \mathcal{F}$ and $\mathcal{F}$ is nonempty.

>>> STEP 2

The family $\mathcal{F}$ is a normal family. In fact, the elements of $\mathcal{F}$ are holomorphic, and they are all bounded by 1. Montel's theorem thus applies, and the family is normal.

>>> STEP 3

Let us define $M = \sup\{|\sigma'(P)| : \sigma \in \mathcal{F}\}$. We claim that M is finite. In fact, let $\bar{D}(P,r)$ be a closed disc about P which is contained in U. Then the Cauchy estimates apply: for any $\sigma \in \mathcal{F}$, we have that

$$|\sigma'(P)| \leq \frac{1!}{r} = \frac{1}{r}.$$

Hence M does not exceed $1/r$.

>>> STEP 4

We claim that there is a function $\sigma_0 \in \mathcal{F}$ such that $\sigma_0'(P) = M$. This is where normal families come in. The present step is of historical interest because Riemann asserted it without proof. Thus his theorem was doubted for several years until the step was corrected using the Dirichlet principle. Nowadays normal families are more commonly used to verify the assertion. The point is that it is fallacious to formulate an extremal problem and simply to assume that it has a solution.

To see how the argument works, we use the definition of supremum to conclude that there is a sequence $\{\sigma_j\}$ in $\mathcal{F}$ such that $|\sigma_j'(P)| \to M$. But since, by Step 2, $\mathcal{F}$ is a normal family, we know that there is a subsequence $\{\sigma_{j_k}\}$ which converges on compact subsets to a limit function σ_0. Now the Cauchy estimates imply that $\{|\sigma_{j_k}'(P)|\}$ converges to M. Hence $|\sigma_0'(P)| = M$. Multiplying σ_0 by a unimodular constant now gives the function that we seek.

>>> STEP 5

We claim that σ_0 is one-to-one on U. In fact, this follows from a clever application of the argument principle. Fix distinct points Q and R in U. Let $0 < s < |Q - R|$. Consider the functions $\psi_k(z) \equiv \sigma_{j_k}(z) - \sigma_{j_k}(Q)$ on $\bar{D}(R,s)$. Since the σ_j are all one-to-one, the functions ψ_k are all nonvanishing on $\bar{D}(R,s)$. By Hurwitz's theorem the limit function $\sigma_0(z) - \sigma_0(Q)$ is either identically zero or is nonvanishing on $\bar{D}(R,s)$. It certainly is not identically zero, for we verified in Step 1 that $\sigma_0'(P) = M > 0$. Thus $\sigma_0(z) \neq \sigma_0(Q)$ for all $z \in \bar{D}(R,s)$, in particular $\sigma_0(R) \neq \sigma_0(Q)$.

Since Q and R were arbitrary, we find that σ_0 is one-to-one.

>>> STEP 6

We claim that σ_0 maps U *onto* the unit disc. It is in this step that the topology of U comes in. One proves, by an advanced calculus argument, that if F is any nonvanishing holomorphic function on U, then $\log F$ is a well-defined holomorphic function on U. Indeed, one can define this logarithm by the formula

$$\log F(z) = \oint_{\gamma_z} \frac{F'(\zeta)}{F(\zeta)}\, d\zeta,$$

where γ_z is any piecewise C^1 path connecting P to z. The topological hypothesis on U guarantees that this definition of $\log F$ is independent of the path.

Knowing that $\log F$ exists allows us to see immediately that any power of F also exists: indeed, if $\alpha \in \mathbf{C}$, then we define

$$F^\alpha(z) = \exp\big(\alpha \log F(z)\big).$$

Now, seeking a contradiction, assume that σ_0 is not onto $D(0,1)$. Say that a point $\beta \in D(0,1)$ is omitted from the image. We now compose with certain Möbius transformations of the disc so that we may exploit the Schwarz lemma.

Recall the Möbius transformations ϕ_a of Section 2. Our hypothesis implies that $\phi_\beta \circ \sigma_0(\zeta)$ never vanishes on U. Thus, the function

$$\mu(\zeta) = \left(\phi_\beta \circ \sigma_0(\zeta)\right)^{1/2}$$

is a well-defined holomorphic function on U. Finally, we let $\tau = \mu(P)$ and set

$$\nu(\zeta) = \frac{|\mu'(P)|}{\mu'(P)} \cdot \left(\phi_\tau \circ \mu(\zeta)\right).$$

Then $\nu \in \mathcal{F}$. By definition, $\nu(P) = 0$ and a calculation shows that

$$|\nu'(P)| = \frac{1 + |\beta|}{2|\beta|^{1/2}} \cdot M > M.$$

This contradicts the choice of σ_0 as having maximal derivative at P. Thus we conclude that σ_0 is onto.

The function σ_0 is the required conformal equivalence of U with $D(0,1)$. □

4. Isolated Singularities and the Theorems of Picard

We begin with a brief discussion of the types of singularities that a holomorphic function can have. Let F be holomorphic on the set $U = D(P,r) \setminus \{P\}$. Then there are three possible ways that F can behave near P:

(i) F is bounded near P;
(ii) F is unbounded near P, but there exists a $k > 0$ such that $(z-P)^k F(z)$ is bounded near P;
(iii) Neither (i) nor (ii) holds.

It is plain that (i), (ii), (iii) are mutually exclusive and cover all possibilities.

In the first instance, the Riemann removable singularities theorem guarantees that $\alpha \equiv \lim_{z\to P} F(z)$ exists and that the function

$$F(z) = \begin{cases} F(z), & z \in U \\ \alpha, & z = P \end{cases}$$

is holomorphic on $D(P,r)$.

In the second case, it can be shown that

$$\lim_{z\to P} |F(z)| = +\infty.$$

The function F is said to have a *pole* at P. If k is the least positive integer which satisfies (ii), then it turns out that F can be written in the form

$$F(z) = (z-P)^{-k} h(z)$$

with h holomorphic on $D(P,r)$ and nonvanishing near P.

The third case is the most complex, and is the one that we wish to concentrate on here. In this circumstance F is said to have an *essential singularity* at P. We begin by proving the Casorati-Weierstrass theorem, which only begins to suggest the level of complexity involved.

□ **Theorem 1.** *Let F be holomorphic on $U = D(P,r) \setminus \{P\}$ and suppose that F has an essential singularity at P. Then, for any $0 < s < r$, the set $\{F(z) : 0 < |z - P| < s\}$ is dense in the complex plane.*

The theorem says that holomorphic functions have a very special mode of behavior near an isolated singularity: either the function has a finite limit, or an infinite limit, or it takes values arbitrarily close to all complex values.

PROOF OF THEOREM 1.
Suppose that the conclusion is false. Then there is a complex number β and an $r > 0$ such that

$$|F(z) - \beta| > r \quad , \quad \text{for all } z \in U.$$

Define

$$h(z) = \frac{1}{F(z) - \beta}.$$

Then $|h(z)|$ is bounded near P by the constant $1/r$. By the Riemann removable singularities theorem, h can be continued analytically through the point P. If $h(P) \neq 0$, then we have a contradiction since then F could be continued analytically through P. But $h(P) = 0$ would imply that F has a pole at P, and that is also false. The resulting contradiction completes the proof. □

It is sometimes useful to think about the point at ∞ just like any point of the finite complex plane. The map

$$z \mapsto \frac{1}{z}$$

allows us to pass back and forth between neighborhoods of 0 and neighborhoods of ∞. In particular, a function F holomorphic on a set $\{z \in \mathbf{C} : |z| > R\}$ is said to have a removable singularity, a pole, or an essential singularity at ∞ if the function $F(1/z)$ has, respectively, a removable singularity, a pole, or an essential singularity at the origin. We now record some striking properties about singularities at infinity.

□ **Proposition 2.** *If F is entire (holomorphic on all of $\mathbf{C}$) and F has a removable singularity at infinity, then F is constant.*

SKETCH OF PROOF.
By examining $F(1/z)$, we see that F must have a finite limit at ∞. Thus F is bounded. By Liouville's theorem, F is constant. □

□ **Proposition 3.** *If F is entire and F has a pole at ∞, then F is a polynomial.*

SKETCH OF PROOF.
Looking at $F(1/z)$, we find that for some $k > 0$ it holds that $z^k F(1/z)$ is bounded near 0. But then $z^{-k}F(z)$ is bounded near ∞. Let P_k be the kth degree Taylor polynomial of F about zero. Then the function

$$h(z) = \frac{F(z) - P_k(z)}{z^k}$$

is holomorphic on the entire plane and bounded at ∞. Thus, by Liouville, h is constant. It follows that F is a polynomial. □

We learn from these two propositions that if F is entire, then F is either a polynomial or F has an essential singularity at ∞. Since the whole point of holomorphic function theory is to study a class of functions which generalizes the properties of polynomials, it is clearly important to understand essential singularities.

Picard's theorems give very sharp results on the value distribution of entire functions and, more generally, holomorphic functions near an essential singularity. We now state them.

□ **Theorem 4** (Picard's Little Theorem). *If F is a nonconstant entire function, then the image of F contains all complex numbers with the possible exception of one value. In other words, if an entire function omits two values, then it is constant.*

□ **Theorem 5** (Picard's Great Theorem). *Let F be holomorphic on an open set $U = D(P,r) \setminus \{P\}$. If F has an essential singularity at P, then for any $0 < s < r$ it holds that $\{F(z) : 0 < |z - P| < s\}$ contains all complex numbers except possibly one value.*

It should now be clear that the great theorem is a generalization of the little theorem. Let us discuss some examples to put the results in perspective.

If $F(z)$ is a nonconstant polynomial and α is any complex number, then the equation $F(z) = \alpha$ always has a solution, by the fundamental theorem of algebra. Thus F takes *all* complex values. If $F(z) = e^z$, then F takes all complex values except 0. The little theorem allows entire functions to omit one complex value. It does not allow omission of two.

Both of Picard's theorems are quite difficult to prove by classical techniques. The original proof involved the construction of a special holomorphic function from the upper half plane to $\mathbf{C}\setminus\{0,1\}$. This function, known as the *elliptic modular function,* is a powerful tool in function theory. Here we briefly discuss the construction of the modular function and its application to the proof of the little theorem.

Begin with a conformal mapping h of the right half of the region Ω, shown in Figure 1, to the upper half plane, normalized so that $h(0) = 0$, $h(1) = 1$, $h(\infty) = (\infty)$.

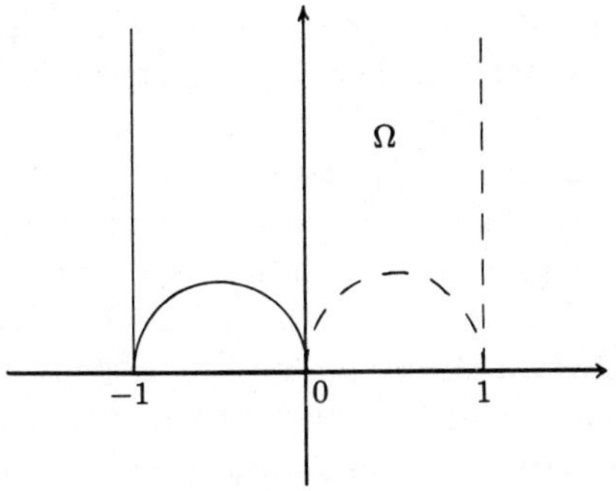

FIGURE 1

By Schwarz reflection, the map h extends to a map from the entire region Ω to $\mathbf{C}\setminus\{0,1\}$. By more delicate reflection arguments, it is possible to continue the function analytically to a map from the entire upper half plane to

$\mathbf{C} \setminus \{0,1\}$. Denote the extended function by the symbol $\lambda(z)$.

Now if F is an entire function whose range omits two values, we may arrange by composition with a linear transformation for those two values to be 0 and 1. Then a germ of $\lambda^{-1} \circ F$, initially defined on a small disc, may be analytically continued to all of $\mathbf{C}$. Thus $\lambda^{-1} \circ F$ is an entire function taking values in the upper half plane. Finally, the holomorphic function

$$\rho(z) = \frac{i - z}{i + z}$$

maps the upper half plane to the unit disc. In summary, the composition $\rho \circ \lambda^{-1} \circ F$ is an entire function which takes values in the unit disc; in other words, it is bounded. By Liouville's theorem, the composition is constant. Unravelling the composition, we find that F is constant. This completes our sketch of the proof of Picard's little theorem.

The construction of the elliptic modular function is both technical and complicated. The analytic continuation argument that is required to extend a germ of $\lambda^{-1} \circ F$ to all of $\mathbf{C}$ requires a great deal of theory. Other proofs of Picard's theorem arise from Nevanlinna theory, or the value distribution theory of entire functions. One of the benefits of our geometric study of complex function theory will be a rather simple and natural proof of both the little and the great Picard theorems.

CHAPTER 1

BASIC NOTIONS OF DIFFERENTIAL GEOMETRY

0. Introductory Remarks

Differential geometry has developed into one of the most powerful mathematical tools of the twentieth century. It has become an integral part of the theories of differential equations, harmonic analysis, and complex analysis, to name just a few examples.

In spite of their importance, the techniques of geometry have not proliferated as much as they might have because of the *complexity of the language*. The characterization of differential geometry as "that portion of mathematics which is invariant under change of notation" is unfortunately rather accurate.

The best way to learn new mathematics is in the context of what is already familiar. In the milieu of one complex variable, the notions of Riemannian metric, of geodesic, and of curvature become rather simple. The device of tensors and bundles is not necessary. The result is that one can learn the flavor and some of the methodology of differential geometry without being encumbered by its notation and machinery.

In this chapter, therefore, we shall learn some of the basic ideas of differential geometry, but *only in the complex plane*; we shall also learn a few simple applications. In later chapters we shall learn about curvature and more advanced applications.

1. Riemannian Metrics and the Concept of Length

In classical analysis a *metric* is a device for measuring distance. If X is a set then a metric γ for X is a function

$$\gamma: X \times X \longrightarrow \mathbf{R}$$

satisfying, for all $x, y, z \in X$,

(1) $\gamma(x,y) = \gamma(y,x)$
(2) $\gamma(x,y) \geq 0$ and $\gamma(x,y) = 0$ iff $x = y$;
(3) $\gamma(x,y) \leq \gamma(x,z) + \gamma(z,y)$.

The trouble with a metric defined in this generality is that it does not interact well with calculus. What sort of interaction might we wish to see?

Given two points $P, Q \in X$, one would like to consider the *curve of least length* connecting P to Q. Any reasonable construction of such a curve leads to a differential equation, and thus we require that our metric lend itself to differentiation. Yet another consideration is curvature: *in the classical setting curvature is measured by the rate of change of the normal vector field.* The concepts of normal and rate of change lead inexorably to differentiation.

Recall that in multivariable calculus we define the arc length of a continuously differentiable curve $\gamma: [a,b] \to \mathbf{R}^2$ by the formula

$$\ell(\gamma) = \int_a^b |\dot{\gamma}(t)| dt. \qquad (*)$$

In point of fact, most calculus books give a heuristic that leads from a reasonable notion of length to this formula. But, rigorously speaking, $(*)$ must be considered to be the *definition* of arc length.

Notice that the definition $(*)$ hinges on the assumption that we have previously defined $|\dot{\gamma}(t)|$, the length of the tangent vector $\dot{\gamma}(t)$ to γ at t. Riemann's insight into differential geometry was that we can define a new metric by specifying a new way to measure the lengths of tangent vectors. The most interesting examples arise when the method of measuring the length of a tangent vector varies from point to point. Thus we now give the word "metric" a new meaning; of course this new meaning will turn out to be intimately related to the older, more classical one.

Definition 1.

If $\Omega \subseteq \mathbf{C}$ is a domain, then a *metric* on Ω is a continuous function $\rho(z) \geq 0$ in Ω which is twice continuously differentiable on $\{z \in \Omega : \rho(z) > 0\}$. If $z \in \Omega$ and $\xi \in \mathbf{C}$ is a vector then we define the *length* of ξ *at* z to be

$$\| \xi \|_{\rho,z} \equiv \rho(z) \cdot |\xi| \,,$$

where $|\xi|$ denotes Euclidean length.

Remark.

Usually our metrics ρ will be strictly positive, but it will occasionally be convenient for us to allow a metric to have isolated zeroes. These will arise as zeroes of holomorphic functions. The zeroes of ρ should be thought of as singular points of the metric.

For the record, the metrics we are considering here are a special case of the type of differential metric called *Hermitian*. This terminology need not concern us here. Classical analysts sometimes call these metrics *conformal metrics* and write them in the form $\rho(z)|dz|$. □

Example 1.

Let Ω be any domain in $\mathbf{C}$. Define $\rho(z) = 1$ for all $z \in \Omega$. Notice that this particular metric yields that if $z \in \Omega$ and $\xi \in \mathbf{C}$ then

$$\| \xi \|_{\rho,z} \equiv \rho(z) \cdot |\xi| = |\xi|.$$

In short, this choice of metric gives the standard Euclidean notion of vector length—and this notion is independent of the base point z. This metric is usually called the *Euclidean metric*. □

Example 2.

Let $\Omega = \{z \in \mathbf{C} : |z| < 1\} \equiv D$, the unit disc. Let

$$\rho(z) = \frac{1}{1 - |z|^2}.$$

This is the *Poincaré metric*, which has been used to gain deep insights into complex analysis on the disc. It will receive our detailed attention later on in this book. For now we do some elementary calculations with the Poincaré metric.

For any $\xi \in \mathbf{C}$ we may calculate that

$$\| \xi \|_{\rho,(1/2+0i)} = \rho(1/2+0i) \cdot |\xi| = \frac{4}{3} \cdot |\xi| \; ;$$

$$\| \xi \|_{\rho,0} = \rho(0) \cdot |\xi| = 1 \cdot |\xi| = |\xi| \; ;$$

$$\| \xi \|_{\rho,(0+.9i)} = \rho(.9i) \cdot |\xi| = \frac{1}{.19} |\xi| = (5.2631578) \cdot |\xi|.$$

□

The notion of the length of a vector varying with the base point is in contradistinction to what we learn in calculus. In calculus, a vector has direction and magnitude but *not* position. Now we declare that a vector *has position* and the way that its magnitude is calculated depends on that position. The next example shows how vectors *with position* arise in practice.

Example 3.

Let $\eta(t) = t$, $0 \le t \le 1/2$ and $\mu(t) = 1/2 + it$, $0 \le t \le 1/2$. Let us calculate

$$\| \dot{\eta}(t) \|_{\rho,\eta(t)} \text{ and } \| \dot{\mu}(t) \|_{\rho,\mu(t)}$$

for the Poincaré metric

$$\rho(z) = \frac{1}{1-|z|^2}.$$

We have $\dot{\eta}(t) = 1$ for all t. Notice that the Euclidean length of $\dot{\eta}(t)$ is 1. However

$$\| \dot{\eta}(t) \|_{\rho,\eta(t)} = \rho(\eta(t)) \cdot |\dot{\eta}(t)| = \frac{1}{1-t^2}.$$

We also have $\dot{\mu}(t) = 1$ for all t and

$$\| \dot{\mu}(t) \|_{\rho,\mu(t)} = \rho(\mu(t)) \cdot |\dot{\mu}(t)| = \frac{1}{3/4-t^2}.$$

□

Tangent vectors will arise for us most often as tangents to curves. It makes sense to think of the tangent vector $\dot{\eta}(t)$ as living at $\eta(t)$ and $\dot{\mu}(t)$ as living at $\mu(t)$. See Figure 1.

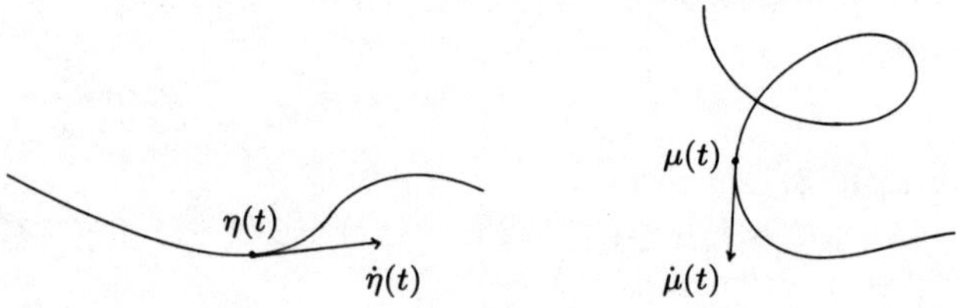

FIGURE 1

Definition 2.

Let $\Omega \subseteq \mathbf{C}$ be a domain and ρ a metric on Ω. If

$$\gamma : [a,b] \to \Omega$$

is a continuously differentiable curve then we define its *length in the metric* ρ to be

$$\ell_\rho(\gamma) = \int_a^b \| \dot{\gamma}(t) \|_{\rho,\gamma(t)} \, dt.$$

The length of a piecewise continuously differentiable curve is defined to be the sum of the lengths of its continuously differentiable pieces.

Notice that this definition is in complete analogy with the classical Euclidean (calculus) definition of arc length. However we are now using an integrand which contains a more general notion of vector magnitude.

Observe also that it follows from the change of variables formula that the length of a curve is independent of its parametrization.

Classical sources write the arc length as

$$\ell_\rho(\gamma) = \int_\gamma \rho(z)|dz| \, ,$$

but we shall not use this notation.

Example 4.

Let $D \subseteq \mathbf{C}$ be the unit disc and let $\rho(z) = 1/(1-|z|^2)$, the Poincaré metric on D. Fix $\epsilon > 0$. Let us calculate the length of the curve $\gamma(t) = t$, $0 \le t \le 1-\epsilon$. Now

$$
\begin{aligned}
\ell_\rho(\gamma) &= \int_0^{1-\epsilon} \| \dot\gamma(t) \|_{\rho,\gamma(t)}\, dt \\
&= \int_0^{1-\epsilon} \frac{|\dot\gamma(t)|}{1-|\gamma(t)|^2} dt \\
&= \int_0^{1-\epsilon} \frac{1}{1-t^2} dt \\
&= \frac{1}{2}\ln\left[\frac{2-\epsilon}{\epsilon}\right].
\end{aligned}
$$

Observe that, when ϵ is small, $\ell_\rho(\gamma)$ is large. In fact

$$\lim_{\epsilon\to 0^+} \ell_\rho(\gamma) = +\infty\ .$$

This suggests that the boundary ∂D is infinitely far from the origin, at least along this particular path γ, in this the Poincaré metric. □

Remark.

Notice that if we consider the metric $\rho(z) \equiv 1$, as in Example 1, then the length of a curve turns out to be the ordinary Euclidean notion of length. □

In a rigorous course on Riemannian geometry one proves that, for a reasonable metric (the metric needs to be *complete*), curves of least length always exist (the proper language for formulating this fact is that of geodesics, curves which have least length in an infinitesimal sense). In the present concrete situation, we may bypass abstractions like this and still come up with useful information:

Example 5.

Equip the disc D with the Poincaré metric $\rho(z) = 1/(1-|z|^2)$. Fix $\epsilon > 0$. Let us prove that among all continuously differentiable curves of the form

$$\mu(t) = t + iw(t) \ , \qquad 0 \le t \le 1-\epsilon \ ,$$

which satisfy $\mu(0) = 0$ and $\mu(1-\epsilon) = 1-\epsilon+0i$, the one of least length is $\gamma(t) = t$. Here $w(t)$ is a continuously differentiable real-valued function.

In fact for any such candidate μ we have

$$\begin{aligned} \ell_\rho(\mu) &= \int_0^{1-\epsilon} \| \dot{\mu}(t) \|_{\rho,\mu(t)} \, dt \\ &= \int_0^{1-\epsilon} \frac{1}{1-|\mu(t)|^2} |\dot{\mu}(t)| dt \\ &= \int_0^{1-\epsilon} \frac{1}{1-t^2-(w(t))^2} \cdot (1+[w'(t)]^2)^{1/2} dt. \end{aligned}$$

However

$$\frac{1}{1-t^2-(w(t))^2} \ge \frac{1}{1-t^2} \quad \text{and} \quad (1+[w'(t)]^2)^{1/2} \ge 1.$$

We conclude that

$$\ell_\rho(\mu) \geq \int_0^{1-\epsilon} \frac{1}{1-t^2}\,dt = \ell_\rho(\gamma).$$

This is the desired result.

Notice that with only small modifications this argument can also be applied to *piecewise* continuously differentiable curves $t + iw(t)$. □

In fact if a piecewise continuously differentiable curve connecting the point $0 \in D$ to $(1-\epsilon)+0i \in D$ is *not* of the form

$$\mu(t) = t + iw(t), \qquad (*)$$

then by elementary comparisons it will be longer than a curve of the form (*) (see Figure 2). We may conclude that the curve γ in the example is the shortest of all curves connecting 0 to $(1-\epsilon)+0i$.

The preceding discussion suggests that if a metric ρ is given on a planar domain Ω, and if P, Q are elements of Ω, then the distance in the metric ρ from P to Q should be defined as follows: Define $\mathcal{C}_\Omega(P,Q)$ to be the collection of all piecewise continuously differentiable curves $\gamma\colon[0,1] \to \Omega$ such that $\gamma(0) = P$ and $\gamma(1) = Q$. Now define the ρ-metric distance from P to Q to be

$$d_\rho(P,Q) = \inf\{\ell_\rho(\gamma) : \gamma \in \mathcal{C}_\Omega(P,Q)\}.$$

Check for yourself that the resulting notion of distance satisfies the metric axioms listed at the outset of this section.

There is some subtlety connected with defining distance in this fashion. If $\rho(z) \equiv 1$, the Euclidean metric, and if Ω is the entire plane, then $d_\rho(P,Q)$ is the ordinary Euclidean distance from P to Q.

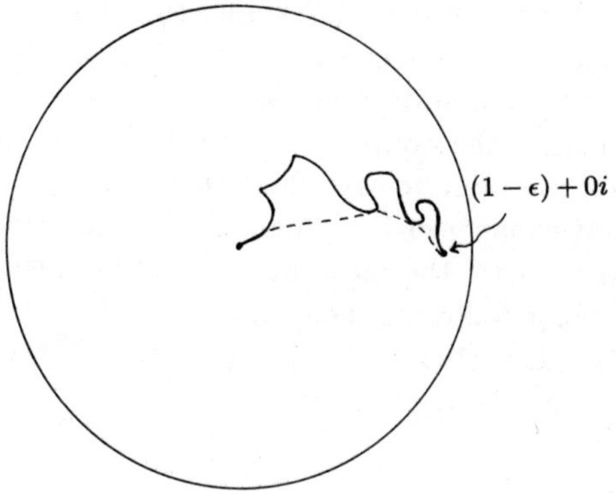

FIGURE 2

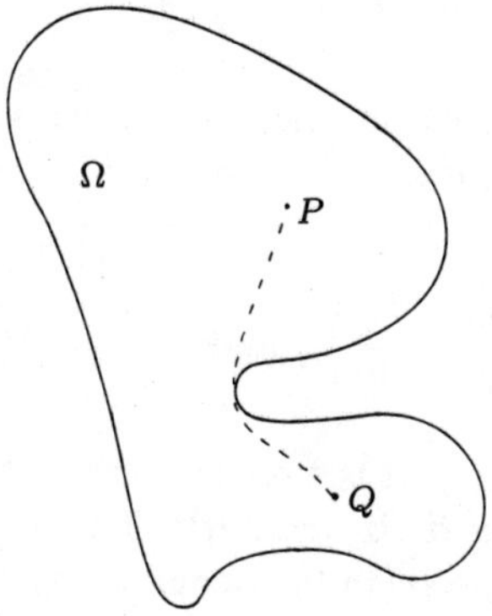

FIGURE 3

But if Ω and P and Q are as shown in Figure 3, then there is no shortest curve connecting P to Q. The *distance* from P to Q is suggested by the dotted curve, but notice that that curve does not lie in Ω. The crucial issue here is whether the domain is complete in the metric, and we shall have more to say about this later.

We may use the language and notation of distance to summarize the meaning of Examples 4 and 5 and the discussion following them: If $\rho(z) = 1/(1-|z|^2)$ is a metric on the disc D, $P = 0$, and $Q = R + i0$, then $d_\rho(P,Q) = (1/2)\log((1+R)/(1-R))$.

2. Calculus in the Complex Domain

In order that we may be able to do calculus computations easily and efficiently in the context of complex analysis, we recast some of the basic ideas in new notation. We define the differential operators

$$\frac{\partial}{\partial z} = \frac{1}{2}\left(\frac{\partial}{\partial x} - i\frac{\partial}{\partial y}\right) \quad \text{and} \quad \frac{\partial}{\partial \bar{z}} = \frac{1}{2}\left(\frac{\partial}{\partial x} + i\frac{\partial}{\partial y}\right).$$

In complex analysis it is more convenient to use these operators than to use $\partial/\partial x$ and $\partial/\partial y$. The reason is as follows. Let $f(z) = u(z) + iv(z)$ be a complex-valued continuously differentiable function (in the sense of multivariable calculus) on a planar domain Ω. Notice that

$$\frac{\partial}{\partial \bar{z}} f(z) = 0 \quad \text{on} \quad \Omega$$

if and only if

$$\frac{1}{2}\left(\frac{\partial}{\partial x} + i\frac{\partial}{\partial y}\right)\left(u(z) + iv(z)\right) = 0 \quad \text{on} \quad \Omega\,.$$

Taking real and imaginary parts of this last equation leads to the pair of equations

$$\frac{\partial}{\partial x}u = \frac{\partial}{\partial y}v \qquad \text{and} \qquad \frac{\partial}{\partial x}v = -\frac{\partial}{\partial y}u.$$

But these are just the Cauchy-Riemann equations. We conclude that

$$\frac{\partial}{\partial \bar{z}}f = 0 \;\text{ on }\; \Omega \;\text{ iff }\; f \;\text{ is holomorphic on }\; \Omega.$$

Further notice that

$$\frac{\partial}{\partial z}z = 1 \qquad \frac{\partial}{\partial z}\bar{z} = 0$$

$$\frac{\partial}{\partial \bar{z}}z = 0 \qquad \frac{\partial}{\partial \bar{z}}\bar{z} = 1.$$

Thus $\partial/\partial z$ and $\partial/\partial \bar{z}$ fit in a natural way into complex function theory. Next observe that if f is holomorphic then

$$\begin{aligned} f'(z) &= \lim_{\mathbf{C}\ni h\to 0} \frac{f(z+h)-f(z)}{h} \\ &= \lim_{\mathbf{R}\ni s\to 0} \frac{f(z+s)-f(z)}{s} \\ &= \frac{\partial f}{\partial x} \\ &= -i\frac{\partial f}{\partial y} \\ &= \frac{\partial f}{\partial z}. \end{aligned}$$

Thus we see that the notion of complex derivative, introduced in Section 0.1, is for holomorphic functions the same as $\partial f/\partial z$.

Finally observe that the Laplace operator

$$\Delta = \frac{\partial^2}{\partial x^2} + \frac{\partial^2}{\partial y^2}$$

may now be written as

$$\Delta = 4\frac{\partial}{\partial z}\frac{\partial}{\partial \bar{z}} = 4\frac{\partial}{\partial \bar{z}}\frac{\partial}{\partial z}.$$

If we are going to use complex derivatives, then we need the usual tools associated to derivatives. The linear properties (that is, the sum and scalar multiplication rules) of $\partial/\partial z$ and $\partial/\partial \bar{z}$ are obvious. The product rule is a bit more tedious, but is safely left as an exercise. However the chain rule is more subtle; it now takes the following form:

□ **Proposition 1.** *If f and g are continuously differentiable functions, and if $f \circ g$ is well defined on some open set $U \subseteq \mathbf{C}$, then we have*

$$\frac{\partial}{\partial z}(f \circ g)(z) = \frac{\partial f}{\partial z}(g(z))\frac{\partial g}{\partial z}(z) + \frac{\partial f}{\partial \bar{z}}(g(z))\frac{\partial \bar{g}}{\partial z}(z)$$

and

$$\frac{\partial}{\partial \bar{z}}(f \circ g)(z) = \frac{\partial f}{\partial z}(g(z))\frac{\partial g}{\partial \bar{z}}(z) + \frac{\partial f}{\partial \bar{z}}(g(z))\frac{\partial \bar{g}}{\partial \bar{z}}(z).$$

PROOF.
We will sketch the proof of the first identity and leave the second as an exercise.

We have

$$\frac{\partial}{\partial z}(f \circ g) = \frac{1}{2}\left(\frac{\partial}{\partial x} - i\frac{\partial}{\partial y}\right)(f \circ g).$$

We write $g(z) = \alpha(z) + i\beta(z)$, with α and β real-valued functions, and apply the usual calculus chain rule for $\partial/\partial x$ and $\partial/\partial y$. We obtain that the last line equals

$$\frac{1}{2}\left(\frac{\partial f}{\partial x}\frac{\partial \alpha}{\partial x} + \frac{\partial f}{\partial y}\frac{\partial \beta}{\partial x} - i\frac{\partial f}{\partial x}\frac{\partial \alpha}{\partial y} - i\frac{\partial f}{\partial y}\frac{\partial \beta}{\partial y}\right). \qquad (*)$$

Now, using the identities

$$\frac{\partial}{\partial x} = \frac{\partial}{\partial z} + \frac{\partial}{\partial \bar{z}} \qquad \text{and} \qquad \frac{\partial}{\partial y} = i\left(\frac{\partial}{\partial z} - \frac{\partial}{\partial \bar{z}}\right),$$

the expression (∗) may, after some tedious calculations, be reduced to the desired formula. □

□ **Corollary 1.1.** *If either f or g is holomorphic then*

$$\frac{\partial}{\partial z}(f \circ g)(z) = \frac{\partial f}{\partial z}(g(z))\frac{\partial g}{\partial z}(z).$$

Here is an example of the utility of this notation.

Example 1.

Let f be a nonvanishing holomorphic function on a planar domain Ω. Then

$$\Delta\big(\log(|f|^2)\big) = 0.$$

In other words, $\log(|f|^2)$ is harmonic.

To see this, fix $P \in \Omega$ and let $U \subseteq \Omega$ be a neighborhood of P on which f has a holomorphic logarithm. Then on U we have

$$\begin{aligned}\Delta\big(\log(|f|^2)\big) &= \Delta\big(\log f + \log \bar{f}\big)\\ &= 4\frac{\partial}{\partial z}\frac{\partial}{\partial \bar{z}}\log f + 4\frac{\partial}{\partial \bar{z}}\frac{\partial}{\partial z}\log \bar{f}\\ &= 0.\end{aligned}$$

□

3. Isometries

In any mathematical subject there are morphisms: functions which preserve the relevant properties being studied. In linear algebra these are linear maps, in Euclidean geometry these are rigid motions, and in Riemannian geometry these are "isometries." We now define the concept of isometry.

Definition 1.

Let Ω_1 and Ω_2 be planar domains and let

$$f : \Omega_1 \to \Omega_2$$

be a continuously differentiable function with only isolated zeroes. Assume that Ω_2 is equipped with a metric ρ. We define the *pullback* of the metric ρ under the map f to be the metric on Ω_1 given by

$$f^*\rho(z) = \rho(f(z)) \cdot \left|\frac{\partial f}{\partial z}\right|.$$

Remark.

The particular form that we use to define the pullback is motivated by the way that f induces mappings on tangent and cotangent vectors, but this motivation is irrelevant for us here.

It should be noted that the pullback of any metric under a conjugate holomorphic f will be the zero metric.

Thus we have designed our definition of pullback so that holomorphic pullbacks will be the ones of greatest interest. This assertion will be made substantive in Proposition 3 below. □

Definition 2.

Let Ω_1, Ω_2 be planar domains equipped with metrics ρ_1 and ρ_2, respectively. Let

$$f : \Omega_1 \to \Omega_2$$

be a one-to-one, onto, continuously differentiable map. If

$$f^* \rho_2(z) = \rho_1(z)$$

for all $z \in \Omega_1$ then f is called an *isometry* of the pair (Ω_1, ρ_1) with the pair (Ω_2, ρ_2).

The differential definition of isometry is very natural from the point of view of differential geometry, but it is not intuitive. The next proposition relates the notion of isometry to more familiar ideas.

□ **Proposition 3.** *Let Ω_1, Ω_2 be domains and ρ_1, ρ_2 be metrics on these respective domains. If*

$$f : \Omega_1 \to \Omega_2$$

is holomorphic and is an isometry of (Ω_1, ρ_1) to (Ω_2, ρ_2) then the following three properties hold:

a) *If $\gamma : [a, b] \to \Omega_1$ is a continuously differentiable curve then so is $f_* \gamma \equiv f \circ \gamma$ and*

$$\ell_{\rho_1}(\gamma) = \ell_{\rho_2}(f_*\gamma).$$

b) *If P,Q are elements of* Ω_1 *then*

$$d_{\rho_1}(P,Q) = d_{\rho_2}(f(P), f(Q)).$$

c) f^{-1} *is also an isometry.*

PROOF.
Assertion (b) is an immediate consequence of (a). Also (c) is a formal exercise in definition chasing. Therefore we shall prove (a).

By definition,

$$\begin{aligned} \ell_{\rho_2}(f_*\gamma) &= \int_a^b \|(f_*\gamma)'(t)\|_{\rho_2, f_*\gamma(t)}\, dt \\ &= \int_a^b \left\| \frac{\partial f}{\partial z}(\gamma(t)) \cdot \dot{\gamma}(t) \right\|_{\rho_2, f_*\gamma(t)} dt. \end{aligned}$$

Now, using the chain rule from the last section, we see that the integrand equals

$$\begin{aligned} \left| \frac{\partial f}{\partial z}(\gamma(t)) \right| \cdot \|\dot{\gamma}(t)\|_{\rho_2, f_*\gamma(t)} &= \left| \frac{\partial f}{\partial z}(\gamma(t)) \right| \cdot |\dot{\gamma}(t)| \cdot \rho_2(f(\gamma(t)) \\ &= \|\dot{\gamma}(t)\|_{f^*\rho_2, \gamma(t)} \\ &= \|\dot{\gamma}(t)\|_{\rho_1, \gamma(t)}. \end{aligned}$$

since f is an isometry. Substituting this back into the formula for the length of $f_*\gamma$ gives

$$\ell_{\rho_2}(f_*\gamma) = \int_a^b \|\dot{\gamma}(t)\|_{\rho_1,\gamma(t)}\, dt = \ell_{\rho_1}(\gamma).$$

This ends the proof. □

It is not difficult to see that any positively oriented rigid motion of the Euclidean plane—rotations, translations, and their compositions (but *not* reflections, because they are conjugate holomorphic)—is an isometry of the Euclidean metric. In the next section we shall consider the isometries of the Poincaré metric on the disc.

If you have previously studied metric space theory or Banach space theory then you may have already encountered the term "isometry." The essential notion is that an isometry should preserve distance. In fact you can prove as an exercise that if f is a holomorphic mapping of (Ω_1, ρ_1) onto (Ω_2, ρ_2) which preserves distance then f is an isometry according to Definition 2. (Hint: compose f with a curve and differentiate.)

4. The Poincaré Metric

The Poincaré metric on the disc has occurred in many of the examples in previous sections. This metric is the paradigm for much of what we want to do in this book, and we want to treat it in some detail here. The Poincaré metric on the disc D is given by

$$\rho(z) = \frac{1}{1 - |z|^2}.$$

(For the record, we note that there is no agreement in the literature as to what constant goes in the numerator; many references use 2.)

In this and succeeding sections, we shall use the phrase "conformal map" to refer to a holomorphic mapping of one planar region to another which is both one-to-one and onto.

□ **Proposition 1.** *Let ρ be the Poincaré metric on the disc D. Let $h : D \to D$ be a conformal self-map of the disc. Then h is an isometry of the pair (D,ρ) with the pair (D,ρ).*

PROOF.
We have that

$$h^*\rho(z) = \rho(h(z)) \cdot |h'(z)|.$$

Because of Theorem 3 of Section 0.2, we now have two cases:

(i) If h is a rotation then $h(z) = \mu \cdot z$ for some unimodular constant $\mu \in \mathbf{C}$. So $|h'(z)| = 1$ and

$$h^*\rho(z) = \rho(h(z)) = \rho(\mu z) = \frac{1}{1-|\mu z|^2} = \frac{1}{1-|z|^2} = \rho(z)$$

as desired.

(ii) If h is a Möbius transformation then

$$h(z) = \frac{z-a}{1-\bar{a}z}, \qquad \text{some constant } a \in D.$$

But then

$$|h'(z)| = \frac{1-|a|^2}{|1-\bar{a}z|^2}$$

and

$$\begin{aligned}
h^*\rho(z) &= \rho\left(\frac{z-a}{1-\bar{a}z}\right) \cdot |h'(z)| \\
&= \frac{1}{1-\left|\frac{z-a}{1-\bar{a}z}\right|^2} \cdot \frac{1-|a|^2}{|1-\bar{a}z|^2} \\
&= \frac{1-|a|^2}{|1-\bar{a}z|^2 - |z-a|^2} \\
&= \frac{1-|a|^2}{1-|z|^2-|a|^2+|a|^2|z|^2} \\
&= \rho(z).
\end{aligned}$$

Since any conformal self-map of D is a composition of maps of the form (i) or (ii), the proposition is proved. □

We know from this result and from Proposition 3 in Section 3 that conformal self-maps of the disc preserve Poincaré distance. To understand what this means, consider the Möbius transformation

$$\phi(z) = \frac{z+a}{1+\bar{a}z} \; ;$$

the function ϕ does not preserve Euclidean distance, but it does preserve Poincaré distance. See Figure 1.

We can use what we have learned so far to calculate the Poincaré metric explicitly.

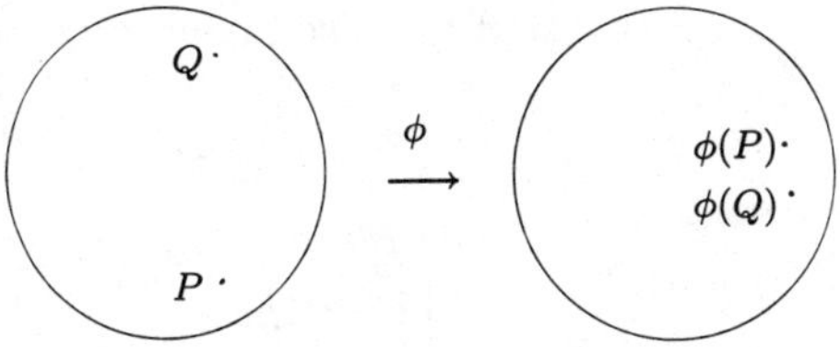

FIGURE 1

□ **Proposition 2.** *If P and Q are points of the disc D then the Poincaré distance of P to Q is*

$$d_\rho(P,Q) = \frac{1}{2}\log\left(\frac{1+\left|\frac{P-Q}{1-\bar{P}Q}\right|}{1-\left|\frac{P-Q}{1-\bar{P}Q}\right|}\right).$$

PROOF.
In case $P = 0$ and $Q = R + i0$ the result was already noted in Section 1. In the general case, note that we may define

$$\phi(z) = \frac{z-P}{1-\bar{P}z},$$

a Möbius transformation of the disc. Then

$$d_\rho(P,Q) = d_\rho(\phi(P),\phi(Q)) = d_\rho(0,\phi(Q)).$$

Next we have

$$d_\rho(0,\phi(Q)) = d_\rho(0,|\phi(Q)|) \qquad (*)$$

since there is a rotation of the disc taking $\phi(Q)$ to $|\phi(Q)|$. Finally,

$$|\phi(Q)| = \left| \frac{P-Q}{1-\bar{P}Q} \right| ,$$

so that $(*)$ together with the special case treated in the the first sentence gives the result. □

One of reasons that the Poincaré metric is so useful is that it induces the same topology as the usual Euclidean metric topology. That is our next result.

□ **Proposition 3.** *The topology induced on the disc by the Poincaré metric is the usual planar topology.*

PROOF.
A neighborhood basis for the topology of the Poincaré metric at the origin is given by the balls

$$B(0,r) = \{z : d_\rho(0,z) < r\}.$$

However a calculation using Proposition 2 yields that these balls are the same as the Euclidean discs

$$\left\{ z : |z| < \frac{e^{2r}-1}{e^{2r}+1} \right\}.$$

These discs form a neighborhood basis for the origin in the Euclidean topology. Thus we find that the two topologies are the same at the origin. Now the origin can be moved to any other point $a \in D$ by the Möbius transformation

$$z \longmapsto \frac{z+a}{1+\bar{a}z}.$$

Since the Poincaré metric is invariant under Möbius transformations, and since Möbius transformations take circles to circles (after all, they are linear fractional), the two topologies are the same at every point. □

One of the most striking facts about the Poincaré metric on the disc is that it turns the disc into a *complete* metric space. How could this be? The boundary is missing! The reason that the disc is complete in the Poincaré metric is the same as the reason that the plane is complete in the Euclidean metric—the boundary is infinitely far away. We now prove this.

□ **Proposition 4.** *The unit disc D, when equipped with the Poincaré metric, is a complete metric space.*

PROOF.
Let p_j be a sequence in D which is Cauchy in the Poincaré metric. Then the sequence is bounded in that metric. So there is a positive, finite M such that

$$d_\rho(0,p_j) \leq M, \qquad \text{all } j.$$

With Proposition 2 this translates to

$$\frac{1}{2}\log\left(\frac{1+|p_j|}{1-|p_j|}\right) \leq M.$$

Solving for $|p_j|$ gives

$$|p_j| \leq \frac{e^{2M} - 1}{e^{2M} + 1} < 1.$$

Thus our sequence is contained in a relatively compact subset of the disc. A similar calculation yields that the sequence must in fact be Cauchy in the Euclidean metric. Therefore it converges to a limit point in the disc, as required for completeness. □

In the proof of Proposition 2 we used the fact, established earlier, that the curve of least length (in the Poincaré metric) connecting 0 to a point of the form $R + i0$ is in fact a Euclidean segment. Let us now calculate the "curve of least length" connecting any two given points P and Q in the disc.

□ **Proposition 5.** *Let P,Q be elements of the unit disc. The "curve of least Poincaré length" connecting P to Q is*

$$\gamma_{P,Q}(t) = \frac{t\frac{Q-P}{1-Q\bar{P}} + P}{1 + t\bar{P} \cdot \frac{Q-P}{1-Q\bar{P}}}, \qquad 0 \leq t \leq 1.$$

PROOF.
Define the Möbius transformation

$$\phi(z) = \frac{z - P}{1 - \bar{P}z}.$$

By what we already know about shortest paths emanating from the origin, the curve $\tau(t) \equiv t \cdot \phi(Q)$ is the shortest curve from $\phi(P) = 0$ to $\phi(Q)$, $0 \le t \le 1$. Applying the isometry ϕ^{-1} to τ we obtain that

$$\phi^{-1} \circ \tau(t) = \frac{(t\phi(Q) + P)}{(1 + t\bar{P}\phi(Q))}$$

is the shortest path from P to Q. Since $\phi^{-1} \circ \tau = \gamma_{P,Q}$, we are done. □

Let us analyze the curves discovered in Proposition 5. First notice that since the curve $\gamma_{P,Q}$ is the image under a linear fractional transformation of a part of a line, the trace of $\gamma_{P,Q}$ is, therefore, either a line segment or an arc of a circle. In fact, if P and Q are collinear with 0, then the formula for $\gamma_{P,Q}$ quickly reduces to that for a segment; otherwise $\gamma_{P,Q}$ traces an arc of a Euclidean circle. Which circle is it?

Matters are simplest if we let t range over the entire real line and look for the whole circle. We find that $t = \infty$ corresponds to the point $1/\bar{P}$. It is now a simple, but tedious, calculation to determine the Euclidean center and radius of the circle determined by the three points $P, Q, 1/\bar{P}$ (note that, by symmetry, the circle will also pass through $1/\bar{Q}$). The circle is depicted in Figure 2. Notice that since the segment $\{t + i0 : -1 \le t \le 1\}$ is orthogonal to ∂D at the endpoints 1 and -1, conformality dictates that the circular arcs of least length provided by Proposition 5 are orthogonal to ∂D at the points of intersection. Some of these "geodesic arcs" are exhibited in Figure 3.

We next see that the Poincaré metric is characterized by its property of invariance under conformal maps.

□ **Proposition 6.** *If $\tilde{\rho}(z)$ is a metric on D which is such that every conformal map of the disc is an isometry of the pair $(D, \tilde{\rho})$ with the pair $(D, \tilde{\rho})$ then $\tilde{\rho}$ is a constant multiple of the Poincaré metric ρ.*

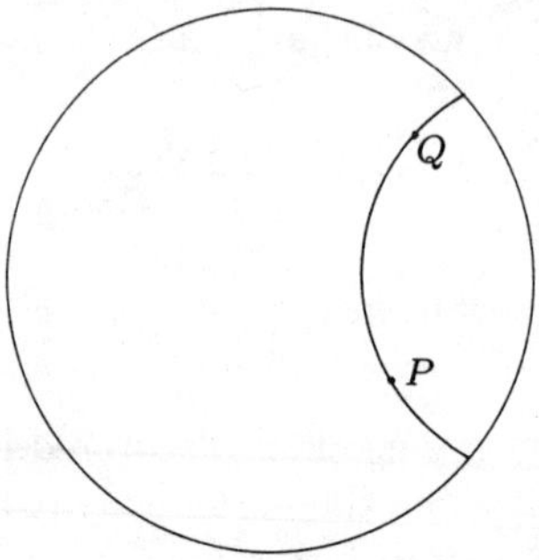

FIGURE 2

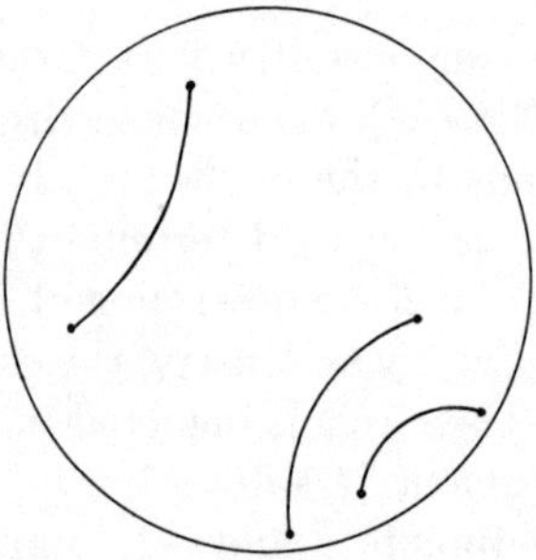

FIGURE 3

PROOF.
The hypothesis guarantees that if $z_0 \in D$ is fixed and

$$h(z) = \frac{z + z_0}{1 + \bar{z}_0 z}$$

then

$$h^*\tilde{\rho}(0) = \tilde{\rho}(0).$$

Writing out the left-hand side gives

$$|h'(0)|\tilde{\rho}(h(0)) = \tilde{\rho}(0)$$

or

$$\tilde{\rho}(z_0) = \frac{1}{1-|z_0|^2} \cdot \tilde{\rho}(0) = \tilde{\rho}(0) \cdot \rho(z_0).$$

Thus we have exhibited $\tilde{\rho}$ as the constant $\tilde{\rho}(0)$ times ρ. □

Now that we know that the Poincaré metric is the right metric for complex analysis on the disc, a natural next question is to determine which other maps preserve the Poincaré metric.

□ **Proposition 7.** *Let $f : D \to D$ be continuously differentiable and let ρ be the Poincaré metric. If f is an isometry of the pair (D,ρ) with the pair (D,ρ) then f is holomorphic. Since f is, by definition of isometry, one-to-one and onto, we may conclude that f is the composition of a Möbius transformation and a rotation* (see Chapter 0).

PROOF.
First suppose that $f(0) = 0$. For $R > 0$ let C_R be the set of points in D which have Poincaré distance R from 0. Since the Poincaré metric is invariant under rotations (after all, rotations are holomorphic self-maps), it follows that C_R is

a Euclidean circle (however this circle will have Euclidean radius $(e^{2R}-1)/(e^{2R}+1)$, not R). Since $f(0)=0$ and f preserves the metric, it follows that $f(C_R)=C_R$. Let $P \in C_R$. Then

$$\frac{|f(P)-f(0)|}{|P-0|}=\frac{|f(P)|}{|P|}=1.$$

Letting $R \to 0^+$ we conclude that f is conformal at the origin (that is, f preserves length infinitesimally).

Now we drop the special hypothesis that $f(0)=0$. Pick an arbitrary $z_0 \in D$ and set $w_0=f(z_0)$. Define

$$\phi(z)=\frac{z+z_0}{1+\bar{z}_0 z}, \qquad \psi(z)=\frac{z-w_0}{1-\bar{w}_0 z}.$$

Also define

$$g=\psi\circ f\circ\phi.$$

Then $g(0)=0$ and g is an isometry (since ψ, f, and ϕ are); therefore the argument in the last paragraph applies and g is conformal at the origin. But ϕ and ψ are conformal at all points. We conclude that f is conformal at z_0. As a result (see [AHL2, pp. 73–74]), f is either holomorphic or conjugate holomorphic. But f is an isometry so that $\partial f/\partial z$ is never 0. It follows that f is holomorphic. □

Propositions 1 and 7 demonstrate that there are plenty of isometries of the unit disc in the Poincaré metric. On the other hand, isometries are very rigid objects. They are completely determined by their first-order behavior at just one point. While a proof of this assertion in general is beyond us at this point, we can certainly prove the result for the Poincaré metric on the disc.

□ **Proposition 8.** *Let ρ be the Poincaré metric on the disc. Let f be a continuously differentiable mapping of the disc to itself. Assume that f is an isometry of the pair (D,ρ) with the pair (D,ρ). If $f(0) = 0$ and $\partial f/\partial z(0) = 1$ then $f(z) \equiv z$.*

PROOF.
By Proposition 7, f must be holomorphic. Since f preserves the origin and is one-to-one and onto, f must be a rotation. Since $f'(0) = 1$, it follows that f is the identity. □

□ **Corollary 8.1.** *Let f and g be isometries of the pair (D,ρ) with the pair (D,ρ). Let $z_0 \in D$ and suppose that $f(z_0) = g(z_0)$ and $(\partial f/\partial z)(z_0) = (\partial g/\partial z)(z_0)$. Then $f(z) \equiv g(z)$.*

PROOF.
We noted in Section 2 that g^{-1} is an isometry. If ψ is a Möbius transformation which takes 0 to z_0 then $\psi^{-1} \circ g^{-1} \circ f \circ \psi$ satisfies the hypothesis of the Proposition. As a result $\psi^{-1} \circ g^{-1} \circ f \circ \psi(z) \equiv z$ or $g(z) \equiv f(z)$. □

5. The Schwarz Lemma

One of the important facts about the Poincaré metric is that it can be used to study not just conformal maps (as in Section 1.4) but all holomorphic maps of the disc. The key to this assertion is the classical Schwarz lemma. We begin with an elegant geometric interpretation of the Schwarz-Pick lemma (see Section 0.2).

□ **Proposition 1.** *Let $f : D \to D$ be holomorphic. Then f is distance decreasing in the Poincaré metric ρ. That is, for any $z \in D$,*

$$f^*\rho(z) \leq \rho(z).$$

The integrated form of this assertion is that if $\gamma : [0,1] \to D$ is a continuously differentiable curve then

$$\ell_\rho(f_*\gamma) \leq \ell_\rho(\gamma).$$

Therefore, if P and Q are elements of D, we may conclude that

$$d_\rho(f(P), f(Q)) \leq d_\rho(P,Q).$$

PROOF.
Now

$$f^*\rho(z) \equiv |f'(z)|\rho(f(z)) = |f'(z)| \cdot \frac{1}{1 - |f(z)|^2}$$

and

$$\rho(z) = \frac{1}{1 - |z|^2}$$

so the asserted inequality is just the Schwarz-Pick lemma. The integrated form of the inequality now follows by the definition of ℓ_ρ and the inequality for the distance d_ρ follows from the definition of distance. □

Notice that in case f is a conformal self-map of the disc then we may apply this corollary to both f and f^{-1} to conclude that f preserves Poincaré distance, giving a second proof of Proposition 1 of Section 4.

We next give an illustration of the utility of the geometric point of view. We will see that the proposition just proved gives a very natural proof of Theorem 2 (see [EAH] for the source of this proof).

□ **Theorem 2.** (Farkas, Ritt) *Let $f : D \to D$ be holomorphic and assume that the image $M = \{f(z) : z \in D\}$ of f has compact closure in D. Then there is a unique point $P \in D$ such that $f(P) = P$. We call P a fixed point for f.*

PROOF.
By hypothesis, there is an $\epsilon > 0$ such that if $m \in M$ and $|z| \geq 1$ then $|m - z| > 2\epsilon$. See Figure 1. Fix $z_0 \in D$ and define

$$g(z) = f(z) + \epsilon(f(z) - f(z_0)).$$

Then g is holomorphic and g still maps D into D. Also

$$g'(z_0) = (1 + \epsilon)f'(z_0).$$

By the preceding corollary, g is distance decreasing in the Poincaré metric. Therefore

$$g^*\rho(z_0) \leq \rho(z_0).$$

Writing out the definition of g^* now yields

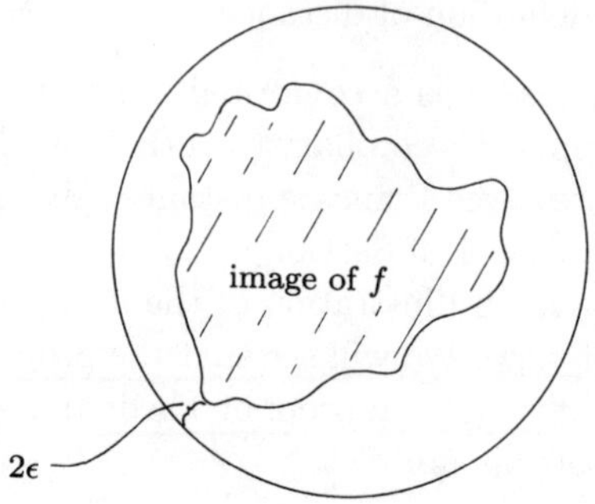

FIGURE 1

$$(1+\epsilon)\cdot f^{*}\rho(z_0) \leq \rho(z_0).$$

Note that this inequality holds for any $z_0 \in D$. But now if $\gamma : [a,b] \to D$ is any continuously differentiable curve then we may conclude that

$$\ell_\rho(f_*\gamma) \leq (1+\epsilon)^{-1}\ell_\rho(\gamma).$$

If P and Q are elements of D and d is Poincaré distance then we have that

$$d(f(P),f(Q)) \leq (1+\epsilon)^{-1}d(P,Q)\ .$$

We see that f is a contraction in the Poincaré metric. Recall that in Section 4 we proved that the disc D is a complete metric space when equipped with the Poincaré metric. By the contraction mapping fixed point theorem (see [LS]), f has a unique fixed point. □

□ **Corollary.** *If f is as in the theorem and P is the unique fixed point then the iterates f, $f \circ f$, $f \circ f \circ f$, ... converge uniformly on compact sets to the constant function P.*

PROOF.
Let f_n denote the nth iterate of f. If we let

$$\bar{B}(P,R) = \{z \in D : d(z,P) \leq R\},$$

then the theorem tells us that

$$f(\bar{B}(P,R)) \subseteq \bar{B}(P,R/(1+\epsilon))$$

and, more generally,

$$f_n(\bar{B}(P,R)) \subseteq \bar{B}(P,R/(1+\epsilon)^n). \qquad (*)$$

Observe that

$$\cup_{j=1}^{\infty} B(P,j) = D.$$

Now Proposition 2 of Section 4 tells us that these non-Euclidean balls are in fact open sets in the usual Euclidean topology. Therefore every (Euclidean) compact subset K of D lies in some $B(P,j)$. Then line $(*)$ implies that

$$f_n(K) \subseteq \bar{B}(P,j/(1+\epsilon)^n).$$

The result follows. □

CHAPTER 2

CURVATURE AND APPLICATIONS

1. Curvature and the Schwarz Lemma Revisited

If $U \subseteq \mathbf{C}$ is a planar domain and ρ is a metric on U, then the *curvature* of the metric ρ at a point $z \in U$ is defined to be

$$\kappa_{U,\rho}(z) = \kappa(z) \equiv \frac{-\Delta \log \rho(z)}{\rho(z)^2}. \qquad (*)$$

(Here zeroes of $\rho(z)$ will result in singularities of the curvature function—κ is undefined at such points.)

Since ρ is twice continuously differentiable, this definition makes sense. It assigns to each $z \in U$ a numerical quantity. The most important preliminary fact about κ is its conformal invariance:

□ **Proposition 1.** *Let U_1 and U_2 be planar domains and $h : U_1 \to U_2$ a conformal map (in particular, h' never vanishes). If ρ is a metric on U_2 then*

$$\kappa_{U_1,h^*\rho}(z) = \kappa_{U_2,\rho}(h(z)), \qquad \forall z \in U_1 .$$

PROOF.
We need to calculate:

$$\begin{aligned}\kappa_{U_1,h^*\rho}(z) &\equiv \frac{-\Delta\log[\rho(h(z))\cdot|h'(z)|]}{[\rho(h(z))\cdot|h'(z)|]^2} \\ &= \frac{-\Delta\log[\rho(h(z))] - \Delta[\log(|h'(z)|)]}{[\rho(h(z))\cdot|h'(z)|]^2}.\end{aligned}$$

Now the second term in the numerator vanishes because of Example 1 in Section 1.2. We may further simplify the numerator to obtain that the last line

$$\begin{aligned}&= \frac{-\left[\Delta\log\rho|_{h(z)}\right]|h'(z)|^2}{[\rho(h(z))\cdot|h'(z)|]^2} \\ &= \frac{-\Delta\log\rho|_{h(z)}}{\rho(h(z))^2} \\ &= \kappa_{U_2,\rho}(h(z)) .\end{aligned}$$

□

Remark.

In fact the proof gives the slightly more general fact that if U_1, U_2 are domains and $f : U_1 \to U_2$ is a holomorphic map (not necessarily one-to-one or onto), then the following holds. If ρ is a metric on U_2, then

$$\kappa_{f^*\rho}(z) = \kappa_\rho(f(z))$$

at every point $z \in U_1$ for which $f'(z) \neq 0$ and $\rho(f(z)) \neq 0$. Usually this is all points except for a discrete set. □

From the point of view of geometry, any differential quantity which is invariant is automatically of great interest. But why do we call κ "curvature"? Why don't we call it "length" or "humidity" or "color"? The answer is that a standard construction in Riemannian geometry assigns to any Riemannian metric a quantity called Gaussian curvature. In the setting of classical Euclidean geometry this quantity coincides with our intuitive notion of what curvature ought to be. In more abstract settings, the structural equations of Cartan lead to a quantity called curvature which is invariant but which has considerably less intuitive content. In the complex plane, the Gaussian curvature calculation leads to the formula which we have used to define κ. One of the pleasant facts about the use of elementary differential geometry in complex function theory is that one does not require all the geometric machinery which leads to the formula for κ in order to make use of κ.

Because of the considerations in the last paragraph, we shall simply accept the definition of κ as given. The geometric calculations *which lead to the definition of* κ are provided in the Appendix. Understanding of the material which follows in the text does not in any way depend on the material in this appendix.

Let us begin our study of curvature by calculating the curvature of the Euclidean metric.

Example 1.

Let U be a planar domain equipped with the Euclidean metric $\rho(z) \equiv 1$. It follows from the definition that $\kappa(z) \equiv 0$. This should be expected, for the Euclidean metric does not change from point to point. □

Example 2.

The metric

$$\sigma(z) = \frac{2}{1 + |z|^2}$$

on $\mathbf{C}$ is called the *spherical metric* (see Section 3 for details). A straightforward calculation shows that the curvature of σ is identically 1. It is a basic theorem of Riemannian geometry that the sphere is the only manifold of constant positive curvature. In fact, the study of manifolds of constant curvature comprises an entire subject in itself (see [WOLF]). □

Next we calculate the curvature of the Poincaré metric on the disc, which calculation will be of great utility for us. Notice that since any point of the disc may be moved to any other by a Möbius transformation, and since curvature is a conformal invariant, we will expect the curvature function to be constant.

□ **Proposition 2.** *Consider the disc D equipped with the Poincaré metric. For any point $z \in D$ it holds that $\kappa(z) = -4$.*

PROOF.
We notice that

$$-\Delta \log \rho(z) = \Delta \log(1 - |z|^2).$$

Now we write $\Delta = 4(\partial/\partial z)(\partial/\partial \bar{z})$ (see Section 1.2) and $|z|^2 = z \cdot \bar{z}$ to obtain that this last

$$= -\frac{4}{(1 - |z|^2)^2} .$$

It follows that $\kappa(z) = -4$. □

The fact that the Poincaré metric on the disc has constant negative curvature turns out to be a special property of the disc. We shall explore this topic later in the book. Now we return to the Schwarz lemma.

It was Ahlfors [AHL1] who first realized that the Schwarz lemma is really an inequality about curvature. In his annotations to his collected works he modestly asserts that "This is an almost trivial fact and anybody who sees the need could prove it at once." However he goes on to say (most correctly) that this point of view has been a decisive influence in modern function theory. It is a good place to begin our understanding of curvature. Here is Ahlfors's version of Schwarz's lemma:

□ **Theorem 3.** *Let the disc $D = D(0,1)$ be equipped with the Poincaré metric ρ and let U be a planar domain equipped with a metric σ. Assume that, at all points of U, σ has curvature not exceeding -4. If $f : D \to U$ is holomorphic then we have*

$$f^*\sigma(z) \le \rho(z), \qquad \forall z \in D.$$

PROOF (following [MIS]).
Let $0 < r < 1$. On the disc $D(0,r)$ define the metric

$$\rho_r(z) = \frac{r}{r^2 - z^2} .$$

Then straightforward calculations (or change of variable) show that ρ_r is the analogue of the Poincaré metric for $D(0,r)$: it has constant curvature -4 and is invariant under conformal maps. Define

$$v = \frac{f^*\sigma}{\rho_r} .$$

Observe that v is continuous and nonnegative on $D(0,r)$ and that $v \to 0$ when $|z| \to r$ (since $f^*\sigma$ is bounded above on $\bar{D}(0,r) \subset\subset D$ while $\rho_r \to \infty$). It follows that v attains a maximum value M at some point $\tau \in D(0,r)$. We will show that $M \leq 1$, hence $v \leq 1$ on $D(0,r)$. Letting $r \to 1^-$ then finishes the proof.

If $f^*\sigma(\tau) = 0$ then $v \equiv 0$. So we may suppose that $f^*\sigma(\tau) > 0$. Therefore $\kappa_{f^*\sigma}$ is defined at τ. By hypothesis,

$$\kappa_{f_*\sigma} \leq -4 .$$

Since $\log v$ has a maximum at τ, we have

$$\begin{aligned} 0 \geq \Delta \log v(\tau) &= \Delta \log f^*\sigma(\tau) - \Delta \log \rho_r(\tau) \\ &= -\kappa_{f_*\sigma}(\tau) \cdot (f^*\sigma(\tau))^2 \\ &\qquad + \kappa_{\rho_r}(\tau) \cdot (\rho_r(\tau))^2 \\ &\geq 4(f^*\sigma(\tau))^2 - 4(\rho_r(\tau))^2 . \end{aligned}$$

This gives

$$\frac{f^*\sigma(\tau)}{\rho_r(\tau)} \leq 1$$

or

$$M \leq 1$$

as desired. □

Observe that the usual Schwarz lemma is a corollary of Ahlfors's version: we take U to be the disc with σ being the Poincaré metric. Let f be a holomorphic mapping of D to $U = D$ such that $f(0) = 0$. Then σ satisfies the hypotheses of the theorem and the conclusion is that

$$f^*\sigma(0) \leq \rho(0).$$

Unravelling the definition of $f^*\sigma$ yields

$$|f'(0)| \cdot \sigma(f(0)) \leq \rho(0).$$

But $\sigma = \rho$ and $f(0) = 0$ so this becomes

$$|f'(0)| \leq 1.$$

Exercise.

The property

$$d(f(P), f(Q)) \leq d(P, Q) ,$$

where d is the Poincaré distance, is called the *distance decreasing property* of the Poincaré metric. Use this property to give a geometric proof of the other inequality in the classical Schwarz lemma. □

With a little more notation we can obtain a more powerful version of the Ahlfors/Schwarz lemma. Let $D(0,\alpha)$ be the open disc of radius α and center 0. For $A > 0$ define the metric ρ_α^A on $D(0,\alpha)$ by

$$\rho_\alpha^A(z) = \frac{2\alpha}{\sqrt{A}(\alpha^2 - |z|^2)}.$$

This metric has constant curvature $-A$. Then we have

□ **Theorem 4.** *Let U be a planar domain which is equipped with a metric σ whose curvature is bounded above by a negative constant $-B$. Then every holomorphic function $f : D(0,\alpha) \to U$ satisfies*

$$f^*\sigma(z) \leq \frac{\sqrt{A}}{\sqrt{B}}\rho_\alpha^A(z), \qquad \forall z \in D(0,\alpha).$$

It is a good exercise to construct a proof of this more general result, modeled of course on the proof of Theorem 3.

In the next three sections we shall see several elegant applications of Theorems 3 and 4.

2. Liouville's Theorem and other Applications

It turns out that curvature gives criteria for when there do or do not exist nonconstant holomorphic functions from a domain U_1 to a domain U_2. The most basic result along these lines is as follows

□ **Theorem 1.** *Let $U \subseteq \mathbf{C}$ be an open set equipped with a metric $\sigma(z)$ having the property that its curvature $\kappa_\sigma(z)$ satisfies*

$$\kappa_\sigma(z) \leq -B < 0$$

for some positive constant B and for all $z \in U$. Then any holomorphic function

$$f : \mathbf{C} \to U$$

must be constant.

PROOF.
For $\alpha > 0$ we consider the mapping

$$f : D(0, \alpha) \to U.$$

Here $D(0, \alpha)$ is the Euclidean disc with center 0 and radius α, equipped with the metric $\rho_\alpha^A(z)$ as in the last section. Fix $A > 0$. Theorem 4 of Section 1 yields, for any fixed z and $\alpha > |z|$, that

$$f^*\sigma(z) \leq \frac{\sqrt{A}}{\sqrt{B}} \rho_\alpha^A(z).$$

Letting $\alpha \to +\infty$ yields

$$f^*\sigma(z) \leq 0$$

hence

$$f^*\sigma(z) = 0.$$

This can only be true if $f'(z) = 0$. Since z was arbitrary, we conclude that f is constant. □

An immediate consequence of Theorem 1 is Liouville's theorem:

□ **Theorem 2.** *Any bounded entire function is constant.*

PROOF.
Let f be such a function. After multiplying f by a constant, we may assume that the range of f lies in the unit disc. However the Poincaré metric on the unit disc has constant curvature -4. Thus Theorem 1 applies and f must be constant. □

Picard's theorem is a dramatic strengthening of Liouville's theorem. It says that the hypothesis "bounded" may be weakened considerably, yet the same conclusion may be drawn.

Let us begin our discussion with a trivial example.

Example 1.
Let f be an entire function taking values in the set

$$S = \mathbf{C} \setminus \{x + i0 : 0 \le x \le 1\}.$$

Following f by the mapping

$$\phi(z) = \frac{z}{z-1},$$

we obtain an entire function $g = \phi \circ f$ taking values in $\mathbf{C}$ less the set $\{x + i0 : x \le 0\}$. If $r(z)$ is the principal branch of the square root function then $h(z) = r \circ g(z)$ is an entire function taking values in the right half plane. Now the Cayley map

$$c(z) = \frac{z-1}{z+1}$$

takes the right half plane to the unit disc. So $u(z) = c \circ h(z)$ is a bounded entire function. We conclude from Theorem 1 that u is constant. Unravelling our construction, we have that f is constant. □

The point of this easy example is that, far from being bounded, an entire function need only omit a segment from its values in order that it be forced to be constant. And a small modification of the proof shows that the segment can be arbitrarily short. Picard considered the question of how small a set the image of a nonconstant entire function can omit.

Let us pursue the same line of inquiry rather modestly by asking whether a nonconstant entire function can omit one complex value. The answer is "yes," for $f(z) = e^z$ assumes all complex values except zero. It also turns out (see Section 3.5) that it is impossible to construct a metric on the plane less a point which has negative curvature bounded away from zero.

The next step is to ask whether a nonconstant entire function f can omit two values. The striking answer, discovered by Picard, is "no." Because of Theorem 1, it suffices for us to prove the following:

□ **Theorem 3.** *Let U be a planar open set such that $\mathbf{C}\setminus U$ contains at least two points. Then U admits a metric μ whose curvature $\kappa_\mu(z)$ satisfies*

$$\kappa_\mu(z) \leq -B < 0$$

for some positive constant B.

PROOF.
After applying an invertible linear map to U we may take the two omitted points to be 0 and 1 (if there are more than two omitted points, we may ignore the extras). Thus we will construct a metric of strictly negative curvature on $\mathbf{C}_{0,1} \equiv \mathbf{C}\setminus\{0,1\}$.

Define

$$\mu(z) = \left[\frac{(1+|z|^{1/3})^{1/2}}{|z|^{5/6}}\right] \cdot \left[\frac{(1+|z-1|^{1/3})^{1/2}}{|z-1|^{5/6}}\right].$$

(After the proof, we will discuss where this nonintuitive definition came from.) Of course μ is positive and smooth on $\mathbf{C}_{0,1}$. We proceed to calculate the curvature of μ.

First notice that, away from the origin,

$$\Delta\left(\log|z|^{5/6}\right) = \frac{5}{12}\Delta\left(\log|z|^2\right) = 0$$

by Example 1 of Section 1.2. Thus

$$\begin{aligned}\Delta\log\left[\frac{(1+|z|^{1/3})^{1/2}}{|z|^{5/6}}\right] &= \frac{1}{2}\Delta\log\left(1+|z|^{1/3}\right)\\ &= 2\frac{\partial}{\partial z\partial\bar{z}}\log\left(1+[z\cdot\bar{z}]^{1/6}\right).\end{aligned}$$

Now a straightforward calculation leads to the identity

$$\Delta\log\left[\frac{(1+|z|^{1/3})^{1/2}}{|z|^{5/6}}\right] = \frac{1}{18}\frac{1}{|z|^{5/3}(1+|z|^{1/3})^2}.$$

The very same calculation shows that

$$\begin{aligned}&\Delta\log\left[\frac{(1+|z-1|^{1/3})^{1/2}}{|z-1|^{5/6}}\right]\\ &\qquad = \frac{1}{18}\frac{1}{|z-1|^{5/3}(1+|z-1|^{1/3})^2}.\end{aligned}$$

The definition of curvature now yields that

$$\begin{aligned}\kappa_\mu(z) = &-\frac{1}{18}\left[\frac{|z-1|^{5/3}}{(1+|z|^{1/3})^3(1+|z-1|^{1/3})}\right.\\ &\left.+\frac{|z|^{5/3}}{(1+|z|^{1/3})(1+|z-1|^{1/3})^3}\right].\end{aligned}$$

We record the following obvious facts:

a) $\kappa_\mu(z) < 0$ for all $z \in \mathbf{C}_{0,1}$;

b) $\lim_{z \to 0} \kappa_\mu(z) = -\frac{1}{36}$;

c) $\lim_{z \to 1} \kappa_\mu(z) = -\frac{1}{36}$;

d) $\lim_{z \to \infty} \kappa_\mu(z) = -\infty$.

It follows immediately that κ_μ is bounded from above by a negative constant. □

Remark.

Now we discuss the motivation for the construction of μ. On looking at the definition of μ, one sees that the first factor is singular at 0 and the second is singular at 1. Let us concentrate on the first of these.

Since the expression defining curvature is rotationally invariant, it is plausible that the metric we define would also be rotationally invariant. Thus it should be a function of $|z|$. Thus one would like to choose an α so that $|z|^\alpha$ defines a metric of negative curvature. However, a calculation reveals that the α which is suitable for z large is not suitable for z small and vice-versa. This explains why the expression has powers both of $|z|$ (for behavior near 0) and of $(1 + |z|)$ (for behavior near ∞). A similar discussion applies to the factors $|z - 1|^\alpha$.

Our calculations thus lead us to design the metric so that it behaves like $|z|^{-4/3}$ near ∞ and behaves like $|z|^{-5/6}$ (respectively, $|z - 1|^{-5/6}$) near 0 (respectively, 1). □

We formulate Picard's little theorem as a corollary of Theorem 1 and Theorem 3:

□ **Corollary 4.** *Let f be an entire analytic function taking values in a set U. If $\mathbf{C} \setminus U$ contains at least two points, then f is constant.*

PROOF.
Since $\mathbf{C} \setminus U$ contains at least two points, Theorem 3 says that there is a metric of strictly negative curvature on U. But then Theorem 1 implies that any entire function taking values in U is constant. □

Entire functions are of two types: there are polynomials and non-polynomials *(transcendental* entire functions). Notice that a polynomial has a pole at infinity. Conversely, any entire function with a pole at infinity is a polynomial (see Section 0.4). So a transcendental entire function cannot have a pole at infinity and, being unbounded (by Liouville), cannot have a removable singularity at infinity. It therefore has an essential singularity at infinity.

Now notice that, by the fundamental theorem of algebra, a polynomial assumes all complex values. For a transcendental function, we analyze its essential singularity by recalling the Casorati-Weierstrass theorem (see Section 0.4): if 0 is an essential singularity for a holomorphic function f on a punctured disc $D'(0,\epsilon) \equiv D(0,\epsilon) \setminus \{0\}$ then f assumes values on $D'(0,\epsilon)$ which are *dense* in the complex plane. One might therefore conjecture that the essential feature of Picard's theorem is not that the function being considered is entire, but rather that it has an essential singularity at infinity. This is indeed the case and is the content of the great Picard theorem, which we treat in Section 4.

3. Normal Families and the Spherical Metric

Recall that in Section 0.3 we briefly considered the concept of normal family. Now we go into the subject more deeply; in particular, we give a more general definition of the concept of normal family. We shall learn that the property of being normal is, when suitably formulated, equivalent to an equicontinuity condition formulated in certain metrics. We begin with our new definition:

Definition 1.

Let Ω be a planar domain and let $\{g_j\}$ be a sequence of complex-valued functions on Ω. We say that the sequence is *normally convergent* to a limit function g if for every $\epsilon > 0$ and every compact set $K \subseteq \Omega$ there is a finite, positive J such that whenever $j > J$ and $z \in K$ then

$$|g_j(z) - g(z)| < \epsilon.$$

In short, the sequence is normally convergent if it converges uniformly on compact subsets of Ω.

Definition 2.

Let Ω be a planar domain and let $\{g_j\}$ be a sequence of complex-valued functions on Ω. We say that the sequence is *compactly divergent* if for every compact set $K \subseteq \Omega$ and every compact set $L \subseteq \mathbf{C}$ there is a finite, positive J such that if $j > J$ and $z \in K$ it holds that $g_j(z) \notin L$.

Briefly, the sequence is compactly divergent if it converges to ∞, uniformly on compact sets.

Definition 3.

Let Ω be a planar domain and let $\mathcal{F}$ be a family of complex-valued functions on Ω. We call $\mathcal{F}$ a *normal family* if every sequence of elements of $\mathcal{F}$ has either a subsequence which is normally convergent or a subsequence which is compactly divergent.

Example 1.

Let $\mathcal{F} = \{f_j\}$, with $f_j(z) = z^j$, $j = 1,2,\ldots$. Then $\mathcal{F}$ is a normal family on $U_1 = \{z : |z| < 1\}$ because every subsequence converges normally to the zero function. Also $\mathcal{F}$ is a normal family on $U_2 = \{z : |z| > 1\}$ because every subsequence is compactly divergent.

However $\mathcal{F}$ is not normal on any domain U_3 which contains a point of the unit circle. The reason is that such a U_3 will contain points inside the circle, on which $\{f_j\}$ converges, and points outside the circle, on which $\{f_j\}$ diverges. □

Notice that, contrary to many complex analysis texts, we do not require in our definition of normal family that the functions in question be holomorphic. In point of fact, normal families are useful outside of complex function theory and it is helpful to have some extra flexibility. However, if we do assume in advance that the functions we are treating are holomorphic, then there are some beautiful theorems about normal families, as we shall now see.

The standard result in the theory of normal families of holomorphic functions is Montel's theorem:

□ **Theorem 4.** *Let Ω be a planar domain and let $\mathcal{F}$ be a family of holomorphic functions on Ω. If for each compact $K \subseteq \Omega$ there is a number M_K such that*

$$|f(z)| \leq M_K, \qquad \forall z \in K, f \in \mathcal{F}, \tag{$*$}$$

then $\mathcal{F}$ is a normal family.

□ **Corollary 4.1.** *If Ω, $\mathcal{F}$ are as in the theorem and if instead of $(*)$ we have a single constant M such that*

$$|f(z)| \leq M, \qquad \forall z \in \Omega,\ f \in \mathcal{F},$$

then $\mathcal{F}$ is a normal family.

Remark.

Notice that no family $\mathcal{F}$ of holomorphic functions satisfying the hypotheses of Montel's theorem could have a compactly divergent subsequence. Thus such a family $\mathcal{F}$ has the property that every sequence of elements of $\mathcal{F}$ has a subsequence which is normally convergent.

After we apply some geometry to the study of normal families, we will be able to derive a version of Montel's theorem which is much more general, and allows compact divergence as well. □

The proof of Montel's theorem is discussed in Section 0.3; further details may be found, for instance in [AHL2, p. 219, ff.]. Now we turn to some examples.

Example 2.

Let $\mathcal{F}$ be the family of all holomorphic functions on a domain Ω which take values in the right half plane. Define

$$\phi(z) = \frac{z-1}{z+1}.$$

Then $\mathcal{G} \equiv \{\phi \circ f : f \in \mathcal{F}\}$ is a family of holomorphic functions taking values in the unit disc. By the corollary to Montel's theorem, $\mathcal{G}$ is a normal family. It follows that $\mathcal{F}$ is a normal family. □

Example 3.

Let

$$U_0 = \mathbf{C} \setminus \{x + i0 : 0 \leq x \leq 1\}.$$

Let $\mathcal{F}$ be all those functions holomorphic in a given domain U and taking values in U_0. We claim that $\mathcal{F}$ is a normal family.

In fact, in Example 1 of Section 2 we found a univalent holomorphic map, call it μ, of U_0 into the unit disc. Therefore $\mathcal{G} = \{\mu \circ f : f \in \mathcal{F}\}$ is a normal family. It follows that $\mathcal{F}$ is a normal family. □

With these examples we are trying to suggest that the concept of normal families is related to Liouville's theorem and Picard's theorem (this is sometimes called "Bloch's principle"). In fact this is true, and we can now begin to see this by looking at matters geometrically. We begin with the concept of the spherical metric.

The stereographic projection p gives a map from the complex plane $\mathbf{C}$ to the Riemann sphere $\hat{\mathbf{C}}$ (see [AHL2, p. 19]). This map is illustrated in Figure 1.

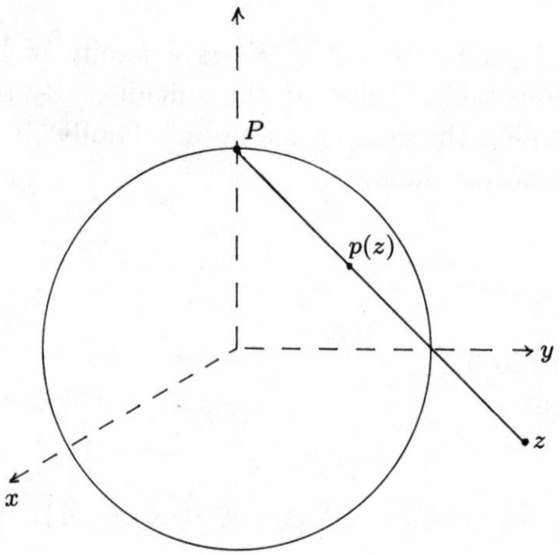

FIGURE 1

We associate to each point z in the complex plane the point $p(z)$ of intersection of the line through z and the "north pole" with the sphere.

If we equip Euclidean 3-space with coordinates as indicated in the figure, then the map p is given explicitly by

$$p(x,y) = \left(\frac{2x}{1+x^2+y^2}, \frac{2y}{1+x^2+y^2}, 1 - \frac{2}{1+x^2+y^2}\right).$$

It is an elegant exercise in differential calculus to check that the metric

$$\sigma(z) = \frac{2}{1+|z|^2}$$

on $\mathbf{C}$ has the following property: if $z, w \in \mathbf{C}$ then $d_\sigma(z,w)$ is precisely the distance (along a great circular arc on the surface of the sphere) from $p(z)$ to $p(w)$. In fact a calculation shows that this distance is

$$d_\sigma(z,w) = 2\arctan\left(\left|\frac{z-w}{1+\bar{w}z}\right|\right). \qquad (*)$$

Differentiation of the formula $(*)$ along a line in the plane gives rise to the definition of σ. (Be reassured that σ can also be thought of as the pullback of the surface Euclidean metric τ on the sphere by the map p. However a precise formulation of this assertion would take us far afield and we omit it.) As an exercise, the reader may wish to derive the formula for $d_\sigma(z,w)$ by imitating our derivation of the formula for Poincaré distance: first calculate the special case $z = 0$, $w = r + 0i$, then use the invariance of the metric under rotations of $\hat{\mathbf{C}}$ (i.e., under linear fractional transformations).

Now suppose that in our study of normal families we study not complex-valued holomorphic functions but meromorphic functions. A meromorphic function is nothing other than a holomorphic function taking values in the Riemann sphere $\hat{\mathbf{C}}$. In order to make this last point clear, let $m(z)$ be a meromorphic function on a domain U with a pole at $P \in U$. Set $I(z) = 1/z$. Then $I \circ m$ is holomorphic in a neighborhood of P. This is precisely the definition of the property "holomorphic" for a $\hat{\mathbf{C}}$-valued function. With this increased generality we can now give a more elegant definition of "normal family." This new definition captures the spirit of the first one, but is strictly more general.

Definition 3′.

A family $\mathcal{F}$ of $\hat{\mathbf{C}}$-valued holomorphic functions on a domain $U \subseteq \mathbf{C}$ is called a *normal family* provided every sequence of elements of $\mathcal{F}$ has a subsequence which is normally convergent.

Remark.

The point of our new definition, and you should prove this as an exercise, is that a compactly divergent sequence (according to Definition 2 above) is actually normally convergent to the point $\infty \in \hat{\mathbf{C}}$. In particular, you should note that if we specialize down to holomorphic functions, Hurwitz's theorem guarantees that any family which is normal according to Definition 3 is also normal according to Definition 3′.

As a second exercise, discuss Example 1 in light of this new definition. □

Now we can give Marty's characterization of normal families in terms of the spherical metric.

□ **Theorem 5.** *Let U be a planar domain and let $\mathcal{F}$ be a family of $\hat{\mathbf{C}}$-valued holomorphic functions on U (equivalently, meromorphic functions). Let τ be the standard Euclidean metric on the surface of the Riemann sphere. The family $\mathcal{F}$ is normal if and only if the family of pullback metrics*

$$\{f^*\tau : f \in \mathcal{F}\}$$

is uniformly bounded on compact subsets of U.

An equivalent, and more concrete, way of saying this is that $\mathcal{F}$ is a normal family if and only if for each compact $K \subseteq U$ there is a constant M_K such that for all $z \in K$ and all $f \in \mathcal{F}$ it holds that

$$\frac{2|f'(z)|}{1+|f(z)|^2} \leq M_K.$$

Remark.

Recall here that if g is a meromorphic function with a pole at P then the derivative of g at P is obtained by differentiating $1/g$; namely, we define

$$g'(P) = \lim_{z \to P} \left(-g^2(z) \cdot (1/g)'(z)\right).$$

□

PROOF.

The equivalence of the two statements follows immediately

from the discussion of the spherical metric and the definition of pullback metric. We now prove the second formulation of the theorem.

First we fix a compact set K and assume that the inequality

$$\frac{2|f'(z)|}{1+|f(z)|^2} \leq M_K, \qquad \forall f \in \mathcal{F},\ z \in K, \qquad (*)$$

holds. Without loss of generality we may assume that K is the closure of a connected open set V. If $\gamma : [a,b] \to V$ is a continuously differentiable curve then the spherical length of $f \circ \gamma$ is

$$\int_a^b \| (f \circ \gamma)'(t) \|_{\tau, f\circ\gamma(t)}\, dt.$$

It follows from $(*)$ that this quantity does not exceed

$$\int_a^b M_K \cdot |\gamma'(t)| dt.$$

But this last is just M_K times the Euclidean length of γ. Notice that this estimate is independent of $f \in \mathcal{F}$. By our definition of distance, it follows that $\mathcal{F}$ is an equicontinuous family of functions when thought of as mapping K equipped with the Euclidean metric to $\hat{\mathbf{C}}$ equipped with the spherical metric. The Ascoli-Arzela theorem (see Section 0.3) now yields immediately that $\mathcal{F}$ is a normal family.

Now we treat the converse direction. Suppose that $\mathcal{F}$ is a normal family. Define the expression

$$f^{\#}(z) = \frac{2|f'(z)|}{1+|f(z)|^2}.$$

Fix a compact set $K \subseteq U$. Seeking a contradiction, we suppose that the functions $\{f^{\#} : f \in \mathcal{F}\}$ are not uniformly bounded on K. Then there is a sequence $f_j \in \mathcal{F}$ such that the maximum of $(f_j)^{\#}$ on K tends to infinity with j. By the normal family hypothesis, we may suppose that each point $P \in U$ has a neighborhood N_P on which $\{f_j\}$ converges normally to a holomorphic function f that takes values in $\hat{\mathbf{C}}$. But then $(f_j)^{\#}$ converges normally on each N_P (it is convenient here to note that $(f_j)^{\#} = (1/f_j)^{\#}$). Since K is compact, we may cover K with finitely many of the sets N_P. It follows that $(f_j)^{\#}$ is bounded on K, contradicting our assumption. □

The thrust of Marty's theorem (and its *proof*) is that the condition of being a normal family is really an equicontinuity condition on the family, *when measured in the correct metric.* That metric becomes plain when the functions are viewed as taking values in $\hat{\mathbf{C}}$.

In the next section we will derive two striking consequences of the work done here.

4. A Generalization of Montel's Theorem and the Great Picard Theorem

Now we reap the reward of our labors. We obtain simple proofs of two of the great results of classical function theory, theorems whose traditional proofs (involving the elliptic modular function, for instance) are quite elaborate. Our main tools here will be the generalized Schwarz lemma and the work in Section 3.

We begin with a more general version (also due to Paul Montel) of the Montel theorem that we discussed earlier:

□ **Theorem 1.** *Let U be a domain in the complex plane. Let P,Q, and R be distinct elements of $\hat{\mathbf{C}}$. Suppose that $\mathcal{F}$ is a family of holomorphic functions taking values in $\hat{\mathbf{C}} \setminus \{P,Q,R\}$. Then $\mathcal{F}$ is a normal family in the sense of Definition 3′ of Section 3.*

PROOF.
We may apply a linear fractional transformation to arrange for $P = 0$, $R = \infty$, and $Q = 1$. Thus we need to show that a family of holomorphic functions taking values in $\hat{\mathbf{C}}_{0,1} \equiv \mathbf{C} \setminus \{0,1\}$ is normal.

Of course it is enough to show that $\mathcal{F}$ is normal on any disc $D(z_0,\alpha) = \{z : |z - z_0| < \alpha\} \subseteq U$. We may assume that $z_0 = 0$.

Let μ be the metric of strictly negative curvature on $\mathbf{C}_{0,1}$ constructed in Theorem 3 of Section 2. After multiplying μ by a positive constant, we may assume that the curvature is bounded from above by -4. By Theorem 4 of Section 1 (the generalized Schwarz lemma), with $A = 4$ and $B = 4$, we have for any $f \in \mathcal{F}$ that

$$f^*\mu(z) \leq \rho_\alpha^A(z), \qquad \forall z \in D(0,\alpha).$$

Now let us compare the spherical metric $\sigma(z)$ restricted to $\mathbf{C}_{0,1}$ with the metric μ. Notice that $\sigma(z)/\mu(z)$ tends to zero as z tends to either 0, 1, or ∞. It follows that there is a constant $M > 0$ such that

$$\sigma(z) \leq M \cdot \mu(z).$$

Therefore we have

$$f^{\#} \equiv f^*\sigma \le M \cdot f^*\mu \le M \cdot f^*\rho_\alpha^A \qquad \text{on } D(0,\alpha).$$

The constant M is independent of $f \in \mathcal{F}$. Thus $f^{\#}$ is bounded on each compact subset of $D(0,\alpha)$, with bound independent of $f \in \mathcal{F}$. Now Marty's theorem implies that $\mathcal{F}$ is a normal family. □

We isolate the first step of the proof as a corollary.

□ **Corollary 1.1.** *If $\mathcal{F}$ is a family of complex-valued holomorphic functions on U all of which omit the same two complex values from their images then $\mathcal{F}$ is a normal family.*

We can now use this generalized Montel theorem to obtain the refinement of Picard's theorem which was promised at the end of Section 2. This result is known as Picard's great theorem.

□ **Theorem 2.** *Let $U = D'(0,\alpha) \equiv D(0,\alpha) \setminus \{0\}$ be a punctured disc and let f be a holomorphic function on U which has an essential singularity at 0. Then f restricted to any deleted neighborhood of 0 omits at most one complex value from its image.*

PROOF.
We prove the contrapositive. After rescaling, we may assume that we have an f on $D' \equiv D(0,1) \setminus 0$ such that the image of f omits the values 0 and 1. We will prove that 0 is therefore either a pole or a removable singularity for f, resulting in a contradiction.

Define $f_n(z) = f(z/n)$, $0 < |z| < 1$, and let $\mathcal{F} = \{f_n\}$. Since all elements of $\mathcal{F}$ take values in $\mathbf{C}_{0,1}$, the family

$\mathcal{F}$ is normal. Because the family $\mathcal{F}$ consists of holomorphic functions, we conclude that there is a subsequence f_{n_k} which either converges normally or diverges compactly.

In the first instance, the subfamily $\{f_{n_k}\}$ is bounded on compact subsets of D'. In particular, it is bounded by some constant M on the set $\{z : |z| = 1/2\}$. But this says that f itself is bounded by M on each of the circles $\{z : |z| = 1/(2n_k)\}$. By the maximum principle, f is bounded by M on $\{z : 0 < |z| < 1/2\}$. So 0 is a removable singularity for f.

In the second case, a similar argument applied to $1/f$ shows that f tends to ∞ as $z \to 0$. Therefore, 0 is a pole for f.

We have shown that if f omits two values in a deleted neighborhood of 0, then f has either a removable singularity or a pole at 0. This completes the proof of the contrapositive of the theorem. □

CHAPTER 3
SOME NEW INVARIANT METRICS

0. Introductory Remarks

Refer to Section 0.3 for the statement and sketch of the proof of the Riemann mapping theorem. The Riemann mapping function is the solution to a certain extremal problem: to find a map of the given domain U into the disc D which is one-to-one, maps a given point P to 0, and has *greatest possible derivative* λ_P at P. The existence of the extremal function, which also turns out to be one-to-one, is established by normal families arguments; the fact that the extremal function is onto is established by an extra argument which is in fact *the only step of the proof where the topological hypotheses about U are used.*

The point of the present discussion is to observe that the scheme we just described can be applied *even* if U is not topologically equivalent to the disc. Constantin Carathéodory's brilliant insight was that the quantity λ_P can be used to construct a metric, now called the Carathéodory metric. We maximize the derivative at P of maps ϕ of U into D such that $\phi(P) = 0$ *but we no longer require the maps to be injective.* Of course the proof of the Riemann mapping theorem will break down at the stage where we attempt to show that the limit map is surjective;

we will also be unable to prove that it is injective. All other steps, *including the existence of the extremal function,* are correct and give rise to a metric (as in Section 3.1 following) on the domain.

A dual construction, using maps $\phi : D \to U$ results in a metric called the Kobayashi or Kobayashi/Royden metric. Both the Carathéodory and the Kobayashi/Royden metrics are motivated by the extremal problem that arises from the Riemann mapping theorem, and are of interest because they endow virtually any domain with a metric. Moreover, these metrics turn out to be invariant under conformal mappings. But the richest benefit is that which can be obtained from studying the interaction of the two metrics.

In this chapter we learn about the Carathéodory and Kobayashi metrics and the insights they lend to the theory of holomorphic mappings. We will also begin to explore what can be learned by comparing the two metrics.

1. The Carathéodory Metric

Fix a domain $U \subseteq \mathbf{C}$. Recall that $D \subseteq \mathbf{C}$ is the unit disc.

Definition 1.

If $P \in U$, define

$$(D,U)_P = \{\text{holomorphic functions } f \text{ from } U \text{ to } D \\ \text{such that } f(P) = 0\}.$$

The *Carathéodory metric* for U at P is defined to be

$$F_C^U(P) = \sup\{|\phi'(P)| : \phi \in (D,U)_P\}.$$

Remark.

As predicted in Section 0, the quantity F_C^U measures, for each P, the extreme value posited in the proof of the Riemann mapping theorem. In that proof, it was only necessary to know that the extreme value existed and was finite. Now we shall gain extra information by comparing this value to other quantities. □

Clearly $F_C^U(P) \geq 0$. (Moreover the Cauchy estimates imply that $F_C^U(P) < \infty$.) Is $F_C^U(P) > 0$ for all P? If U is bounded the answer is yes. For then

$$U \subseteq D(0,R) = \{z \in \mathbf{C} : |z| < R\}$$

for some $R > 0$. But then the map

$$\phi : \zeta \longmapsto \frac{\zeta - P}{2R}$$

satisfies $\phi \in (D,U)_P$. Thus

$$F_C^U(P) \geq |\phi'(P)| = \frac{1}{2R} > 0.$$

However if U is unbounded, then F_C^U may degenerate. If $U = \mathbf{C}$, for instance, then any $f \in (D,U)_P$ is constant hence $F_C^U \equiv 0$. The same holds for U equalling

$\mathbf{C}$ less a discrete set of points (by the Riemann removable singularities theorem). Which domains have nondegenerate Carathéodory metric? Analytic capacity is a device for answering this question (see [GAR]), but we cannot pursue this topic here.

The primary interest of the Carathéodory metric is that it generalizes Proposition 1 in Section 1.5 to arbitrary domains.

□ **Proposition 2.** *Let U_1, U_2 be domains in $\mathbf{C}$. Let ρ_j be the Carathéodory metric on U_j. If $h : U_1 \to U_2$ is holomorphic then h is distance decreasing from (U_1, ρ_1) to (U_2, ρ_2). In other words,*

$$h^*\rho_2(z) \leq \rho_1(z), \qquad \forall z \in U_1.$$

□ **Corollary 2.1.** *If $\gamma : [0,1] \to U_1$ is a piecewise continuously differentiable curve then*

$$\ell_{\rho_2}(h_*\gamma) \leq \ell_{\rho_1}(\gamma).$$

□ **Corollary 2.2.** *If $P_1, P_2 \in U_1$ then*

$$d_{\rho_2}(h(P_1), h(P_2)) \leq d_{\rho_1}(P_1, P_2).$$

□ **Corollary 2.3.** *If h is conformal then h is an isometry of (U_1, ρ_1) to (U_2, ρ_2).*

The corollaries are proved just as they were proved for the Poincaré metric in Chapter 1. We concentrate on proving the Proposition.

PROOF OF PROPOSITION 2.
Fix $P \in U_1$ and set $Q = h(P)$. Notice that if $\phi \in (D, U_2)_Q$ then $\phi \circ h \in (D, U_1)_P$. Thus

$$\begin{aligned} F_C^{U_1}(P) &\geq |(\phi \circ h)'(P)| \\ &= |\phi'(Q)| \cdot |h'(P)|. \end{aligned}$$

Taking the supremum over all $\phi \in (D, U_2)$ yields

$$F_C^{U_1}(P) \geq F_C^{U_2}(Q) \cdot |h'(P)|$$

or

$$\rho_1(P) \geq h^* \rho_2(P).$$

□

Next we calculate the Carathéodory metric for the disc.

□ **Proposition 3.** *The Carathéodory metric on the disc coincides with the Poincaré metric.*

PROOF.
First we calculate the metric at the origin. If $\phi \in (D, D)_0$ then, by the Schwarz lemma, $|\phi'(0)| \leq 1$. Therefore

$$F_C^D(0) \leq 1.$$

But the map

$$\phi(\zeta) = \zeta$$

satisfies $\phi \in (D, D)_0$ and $\phi'(0) = 1$. Thus

$$F_C^D(0) = 1.$$

Because every conformal map of the disc is an isometry of the Carathéodory metric, Proposition 6 of Section 1.4 now implies that the Carathéodory metric and the Poincaré metric are equal. □

Given what we've learned so far, we might suspect that any isometry of the Carathéodory metric must be a conformal map. This is indeed true—in fact a much stronger assertion holds—provided the domains in question have nondegenerate Carathéodory metrics. We postpone this topic until after we have introduced the Kobayashi metric.

2. The Kobayashi Metric

Fix a domain $U \subseteq \mathbf{C}$.

Definition 1.
If $P \in U$, define

$$(U,D)_P = \{\text{holomorphic functions } f \text{ from } D \text{ to } U \text{ such that } f(0) = P\}.$$

The *Kobayashi* (or Kobayashi/Royden) *metric* for U at P is defined to be

$$F_K^U(P) = \inf\left\{\frac{1}{|\phi'(0)|} : \phi \in (U,D)_P\right\}.$$

Remark.

As we remarked in the last section about the Carathéodory metric, the Kobayashi metric quantifies a certain extremal problem and will be useful for comparison purposes. The particular form of the definition of the Kobayashi metric will turn out to facilitate comparisons with the Carathéodory metric; in particular, the two metrics will turn out to interact nicely with the classical Schwarz lemma. □

Clearly $F_K^U(P) \geq 0$. In order to learn more about F_K^U, we now compare it with the Carathéodory metric.

□ **Proposition 2.** *For all $P \in U$,*

$$F_C^U(P) \leq F_K^U(P).$$

PROOF.
Let $\phi \in (D,U)_P$ and $\psi \in (U,D)_P$. Then $\phi \circ \psi : D \to D$ and $\phi \circ \psi(0) = 0$. It follows from Schwarz's lemma that

$$|(\phi \circ \psi)'(0)| \leq 1$$

or

$$|\phi'(P)| \leq \frac{1}{|\psi'(0)|}.$$

Taking the supremum over all ϕ gives

$$F_C^U(P) \leq \frac{1}{|\psi'(0)|}.$$

Now taking the infimum over all ψ yields

$$F_C^U(P) \leq F_K^U(P).$$

□

An immediate consequence of this last result is that if U is bounded then F_K^U is nondegenerate (for then F_C^U is). On the other hand, if $U = \mathbf{C}$ then $F_K^U \equiv 0$; for given $P \in \mathbf{C}$ then

$$\phi_R(\zeta) = P + R\zeta$$

satisfies $\phi_R \in (U,D)_P$, any $R > 0$. Thus

$$F_K^U(P) \leq \frac{1}{|\phi_R'(0)|} = \frac{1}{R} \,;$$

letting $R \to \infty$, we have $F_K^U(P) = 0$.

In analogy with Proposition 2 of the last section we now have:

□ **Proposition 3.** *Let U_1, U_2 be domains equipped with the Kobayashi metric (denoted, respectively, by ρ_1 and ρ_2). If $h : U_1 \to U_2$ is holomorphic then h is distance decreasing from (U_1, ρ_1) to (U_2, ρ_2). That is,*

$$h^*\rho_2(z) \leq \rho_1(z), \qquad \forall z \in U_1.$$

□ **Corollary 3.1.** *If $\gamma : [0,1] \to U_1$ is a piecewise continuously differentiable curve then*

$$\ell_{\rho_2}(h_*\gamma) \leq \ell_{\rho_1}(\gamma).$$

□ **Corollary 3.2.** *If $P_1, P_2 \in U_1$ then*

$$d_{\rho_2}(h(P_1), h(P_2)) \leq d_{\rho_1}(P_1, P_2).$$

□ **Corollary 3.3.** *If h is conformal then h is an isometry of (U_1, ρ_1) to (U_2, ρ_2).*

PROOF OF PROPOSITION 3.
Fix $P \in U_1$ and put $Q = h(P)$. Choose $\phi \in (U_1, D)_P$. Then $h \circ \phi \in (U_2, D)_Q$. We have

$$\begin{aligned} F_K^{U_2}(Q) &\leq \frac{1}{|(h \circ \phi)'(0)|} \\ &= \frac{1}{|h'(P)||\phi'(0)|}. \end{aligned}$$

Taking the infimum over all $\phi \in (U_1, D)_P$ gives

$$F_K^{U_2}(Q) \leq \frac{1}{|h'(P)|} \cdot F_K^{U_1}(P)$$

or

$$(h^* F_K^{U_2})(P) \leq F_K^{U_1}(P).$$

□

Using Proposition 3, we may give another example of domains $U \subseteq \mathbf{C}$ with degenerate Kobayashi metric. Let $\mathbf{C}_0 = \mathbf{C} \setminus \{0\}$. Then

$$h : \mathbf{C} \rightarrow \mathbf{C}_0$$
$$\zeta \rightarrow e^{\zeta}$$

is holomorphic. The map h will be distance decreasing in the Kobayashi metric. Since F_K is identically 0 on $\mathbf{C}$, it follows that F_K is identically zero on $\mathbf{C}_0$.

As an exercise, prove that we cannot continue this line of reasoning to see that F_K^U is degenerate when $U = \mathbf{C} \setminus \{0,1\}$. In fact, we shall later see that F_K^U is nondegenerate on $U \equiv \mathbf{C} \setminus \{P_1, \ldots, P_k\}$ provided the P_j's are distinct and $k \geq 2$ (notice that this situation is in contrast to that for the Carathéodory metric as discussed in Section 1). The matter will be addressed in Section 5; there we will see it connected in an elegant way with normal families—recall the special role of $\mathbf{C}_{0,1}$ in that theory. Meanwhile, we turn attention to more elementary questions regarding the Kobayashi metric.

□ **Proposition 4.** *For $U = D \subseteq \mathbf{C}$, the Kobayashi metric equals the Poincaré metric.*

PROOF.
For $P \in D$ we have

$$F_K^D(P) \geq F_C^D(P) = \rho(P),$$

where ρ is the Poincaré metric on the disc. For an inequality in the opposite direction, first take $P = 0$. Let $\phi \in (D, D)_0$ be given by

$$\phi(\zeta) = \zeta.$$

Then

$$F_K^D(0) \leq \frac{1}{|\phi'(0)|} = 1 = \rho(0).$$

It follows that

$$F_K^D(0) = 1 = \rho(0).$$

By Proposition 3 above and Proposition 6 of Section 1.4, we conclude that

$$F_K^D(P) = \rho(P), \qquad \forall P \in D.$$

□

The first major result of this chapter is a characterization of the disc in terms of the Carathéodory and Kobayashi metrics (at this point have another look at Section 0 to see why this result is a metric version of the Riemann mapping theorem):

□ **Theorem 5.** *Let $U \subseteq \mathbf{C}$ be a domain. The domain U is conformally equivalent to the disc if and only if there is a point $P \in U$ such that*

$$F_C^U(P) = F_K^U(P) \neq 0.$$

PROOF.
If U is conformally equivalent to the disc then let

$$h : U \to D$$

be a conformal map. For any point $P \in U$ we have that

$$F_C^U(P) = (h^* F_C^D)(P)$$

which, by Proposition 2 of Section 1,

$$= (h^* \rho)(P) \ ;$$

but by Proposition 4 this

$$= (h^* F_K^D)(P).$$

By Corollary 3 to Proposition 3 we conclude that this last equals $F_K^U(P)$, proving the easy half of the theorem.

For the converse, choose $\phi_j \in (D,U)_P$ such that

$$|\phi_j'(P)| \to F_C^U(P)$$

and choose $\psi_j \in (U,D)_P$ such that

$$\frac{1}{|\psi_j'(0)|} \to F_K^U(P).$$

Since $\{\phi_j\}$ is uniformly bounded above by 1, we may extract a subsequence $\{\phi_{j_k}\}$ converging to a normal limit ϕ_0. Consider

$$h_{j_k} = \phi_{j_k} \circ \psi_{j_k}.$$

Passing to another subsequence, which we denote by h_{j_ℓ}, we may suppose that h_{j_ℓ} converges normally to a limit h_0. Notice that $h_0(0) = 0$ so that h_0 is not a unimodular constant; therefore h_0 maps D to D. After renumbering, we call this last sequence

$$h_\ell = \phi_\ell \circ \psi_\ell.$$

Recall that when a sequence of holomorphic functions converges normally, then so does the sequence of its derivatives (see the Cauchy estimates in Section 0.1). It follows that

$$\begin{aligned}
|h_0'(0)| &= \lim_{\ell\to\infty} \left|(\phi_\ell \circ \psi_\ell)'(0)\right| \\
&= \lim_{\ell\to\infty} |\phi_\ell'(P)| \cdot |\psi_\ell'(0)| \\
&= F_C^U(P) \cdot \frac{1}{F_K^U(P)} \\
&= 1.
\end{aligned}$$

By Schwarz's lemma, $h_0(\zeta) = \mu \cdot \zeta$ for some unimodular constant μ. Thus we have

$$\mu \cdot \zeta = h_0(\zeta) = \lim_{\ell \to \infty} (\phi_\ell \circ \psi_\ell(\zeta)). \qquad (*)$$

After composing the functions ϕ_ℓ with a rotation, we may assume that $\mu = 1$.

Now $\mathbf{C} \setminus U$ must contain at least two points (else F_K^U would be identically 0, contradicting our hypothesis). Hence, by Theorem 1 of Section 2.4, $\{\psi_\ell\}$ forms a normal family. Say that a subsequence ψ_{ℓ_m} converges to some ψ_0. After renumbering, we rewrite $(*)$ as

$$\begin{aligned} \zeta &= h_0(\zeta) \\ &= \lim_{m \to \infty} (\phi_m \circ \psi_m(\zeta)) \qquad (**) \\ &= \phi_0 \circ \psi_0(\zeta). \end{aligned}$$

Since h_0 is surjective, we conclude that ϕ_0 is surjective.

The function ψ_0 is a nonconstant holomorphic function (since $\psi_0'(0) \neq 0$) hence its image is open. We claim that the image is also closed (in the relative topology of U). To see this, let $\psi_0(\zeta_j)$ be elements of that image converging to a limit point $q \in U$. Since ϕ_0 is continuous, it follows that $\phi_0(\psi_0(\zeta_j))$ converges to a limit point r. But then $(**)$ tells us that $\zeta_j \to r$. It follows from the continuity of ψ_0 that $\psi_0(r) = q$, so q is in the image of ψ_0. We thus have that the image of ψ_0 is both open and closed, and it is nonempty. Since U is connected, it follows that the image of ψ_0 equals U, hence ψ_0 is surjective. Thus, since h_0 is injective, so must ϕ_0 be injective. We conclude that ϕ_0 is the desired conformal map of U to D. □

Notice that the theorem is false if we allow the metrics to degenerate. For example, if $U = \mathbf{C} \setminus 0$ then $F_C^U = F_K^U \equiv 0$ yet U is definitely not conformally equivalent to the disc.

As predicted in the last section, we now prove that the only isometries of the Carathéodory or Kobayashi metrics which fix a point are conformal maps. In fact we shall prove something much stronger.

□ **Theorem 6.** *Let $U \subseteq \mathbf{C}$ be a domain on which the Kobayashi metric is nondegenerate and fix $P \in U$. Suppose that*

$$f : U \longrightarrow U$$

is a holomorphic function such that $f(P) = P$. Assume that f is an isometry of either the Carathéodory or the Kobayashi metric at P; that is, suppose that either

$$f^* F_C^U(P) = F_C^U(P)$$

or

$$f^* F_K^U(P) = F_K^U(P)$$

and that the metric does not vanish at P. Then f is a conformal map of U to U.

Remark.

This theorem is a remarkable rigidity statement: the global condition that f map U to U, together with the differential condition at P, forces f to be one-to-one and onto. □

PROOF OF THE THEOREM.
Since U has nondegenerate Kobayashi metric, $\mathbf{C} \setminus U$ must contain at least two points. This implies that any family of holomorphic functions $\{g_\alpha\}$ taking values in U is normal (by Theorem 1 of Section 2.4). We will use this observation repeatedly.

Now the hypothesis that

$$f^* F_C^U(P) = F_C^U(P) \quad \text{or} \quad f^* F_K^U(P) = F_K^U(P)$$

implies that

$$|f'(P)| = 1.$$

Define

$$\begin{aligned} f^1 &= f \\ f^2 &= f \circ f \\ &\ldots \\ f^j &= f^{j-1} \circ f, \qquad j \geq 2. \end{aligned}$$

Then $\{f^j\}$ is a normal family. Since the numbers $(f^j)'(P)$ all have unit modulus, there is a subsequence $\{f^{j_\ell}\}$ such that $(f^{j_\ell})'(P) \to 1$. (Exercise: if $f'(P)$ has argument which is a rational multiple of π, then the assertion is clear; if it is an irrational multiple of π, then the set of $(f^j)'(P)$ form a dense subset of the circle.) Passing to another subsequence, which we still denote by $\{f^{j_\ell}\}$, we may also suppose that f^{j_ℓ} converges normally to a holomorphic function $\tilde{f}$. It follows that $\tilde{f}'(P) = 1$. We claim in fact that $\tilde{f}(z) \equiv z$.

To prove this, suppose not. Assume for simplicity that $P = 0$. Then, for z near 0,

$$\tilde{f}(z) = 0 + z + \text{ (higher order terms)}.$$

Let $a_m z^m$, $m \geq 2$, be the first nonvanishing power of z which appears after z (since we are assuming that $\tilde{f}$ is not the function z, there must be one). Notice that

$$\begin{aligned} \tilde{f}^2 &\equiv \tilde{f} \circ \tilde{f} = z + 2a_m z^m + \dots \\ \tilde{f}^3 &\equiv \tilde{f} \circ \tilde{f} \circ \tilde{f} = z + 3a_m z^m + \dots \\ &\dots \\ \tilde{f}^k &\equiv \tilde{f} \circ \tilde{f} \circ \dots \circ \tilde{f} = z + k a_m z^m + \dots \end{aligned} \tag{$*$}$$

On the other hand, $\{\tilde{f}^k\}$ forms a normal family, so there is a subsequence $\tilde{f}^{k_q}$ converging normally on U. Hence

$$\left(\frac{\partial}{\partial z}\right)^m \tilde{f}^{k_q}(P)$$

converges. But $(*)$ clearly shows that

$$\left(\frac{\partial}{\partial z}\right)^m \tilde{f}^k(P) = m! \cdot k \cdot a_m$$

blows up with k. This contradiction can only be resolved if $a_m = 0$. Therefore $\tilde{f}(z) \equiv z$.

Now we have that $f^{j_\ell}(z) \to z$ normally. We claim that this implies that f is a conformal map. Indeed, consider the family $\{f^{j_\ell - 1}\}$. It is normal, so there is a subsequence, call it $\{f^{j_r - 1}\}$, which converges normally to a limit g. Notice that g is not constant since $g'(P)$ is not zero. Then

$$\begin{aligned} f \circ g(z) &= f \circ \lim_{r \to \infty} f^{j_r - 1}(z) \\ &= \lim_{r \to \infty} f^{j_r}(z) \\ &\equiv z. \end{aligned}$$

Similarly,

$$g \circ f(z) \equiv z.$$

Thus f is one-to-one and onto. It is certainly holomorphic, so it is a conformal map as claimed. □

It is instructive to view Theorem 6 as a generalization of the uniqueness part of the Schwarz lemma. For simplicity take U to be a bounded domain. Then U is contained in some disc, and a disc has nondegenerate Kobayashi metric. Therefore, by the distance decreasing property of the Kobayashi metric applied to the inclusion map, the domain U also has nondegenerate Kobayashi metric. Suppose that f is a holomorphic function mapping U to U that fixes a point P in U. The distance decreasing property of the Kobayashi metric tells us that $|f'(P)| \leq 1$. Theorem 6 now says that equality obtains if and only if f is one-to-one and onto. □

3. The Completeness of the Carathéodory and Kobayashi Metrics

In this section we prove that if U is a domain with a reasonably nice boundary then it is complete, as a metric space, when equipped with either the Carathéodory or Kobayashi metrics. A warning is in order here: this is simultaneously the richest and the most difficult section of the book; for it uses both the metric language and tricky constructions from analysis. But the section is self-contained, and provides an introduction to a lot of important and beautiful material. Moreover, the payoff for the work, in both this section and the next, is more than adequate compensation for the effort expended.

Now we turn to the treatment of completeness of our two metrics on a reasonable class of domains. It is enough to prove this for the Carathéodory metric, since any sequence which is Cauchy in the Kobayashi metric is also Cauchy in the Carathéodory metric (exercise: provide the details). We begin by defining what we mean by "reasonably nice boundary."

Definition 1.

A curve $\gamma : [a,b] \to \mathbf{C}$ is called a *closed twice continuously differentiable curve* if γ is twice continuously differentiable (with one-sided derivatives as usual at the endpoints), if $\gamma(a) = \gamma(b)$, and if the one-sided derivatives of γ, up to and including order 2, at a equal those at b.

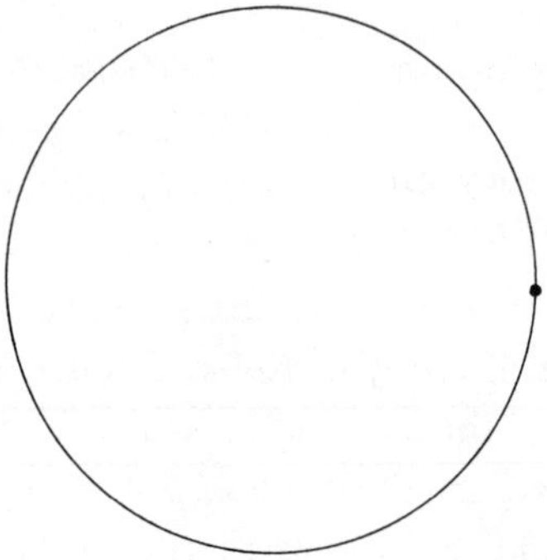

FIGURE 1

Example 1.

The curve $\gamma(t) = e^{it}$, $0 \le t \le 2\pi$, is a closed twice continuously differentiable curve. See Figure 1. Notice that the agreement of the first and second derivatives at the endpoints causes a smooth transition in the figure at the point $1 + i0$ where the two ends meet.

The curve

$$\mu(t) = (\cos 2t \cos t) + i(\cos 2t \sin t), \qquad -\frac{\pi}{4} \le t \le \frac{\pi}{4}$$

is closed, but *not* twice continuously differentiable (because the derivatives at $t = -\pi/4$ and $t = \pi/4$ do not agree). The curve is shown in Figure 2. □

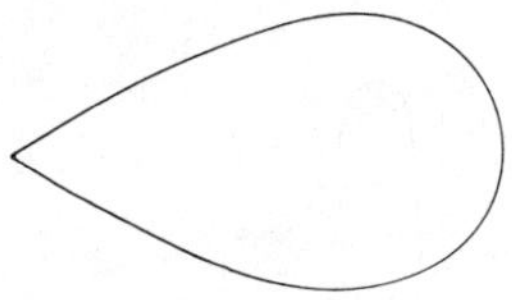

FIGURE 2

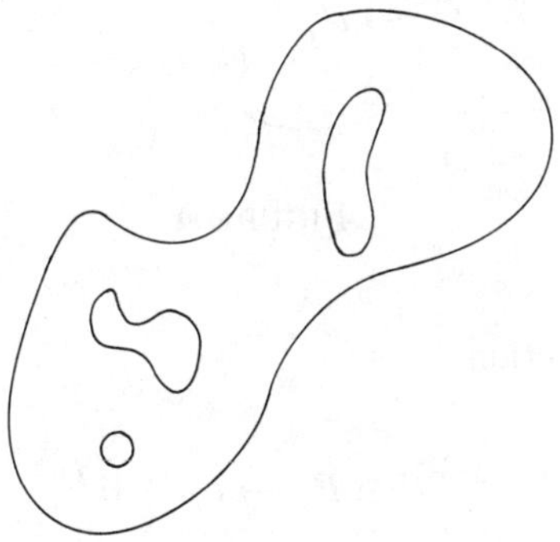

FIGURE 3

Definition 2.

A bounded domain $U \subseteq \mathbf{C}$ is said to have *twice continuously differentiable* (or C^2) boundary if the boundary consists of finitely many pairwise disjoint, simple closed, twice continuously differentiable curves. See Figure 3.

Observe that, by the usual constructions of multivariable calculus, a domain with twice continuously differentiable boundary possesses at each boundary point P a well-defined unit outward normal ν_P and a well-defined unit inward normal ν'_P. See Figure 4.

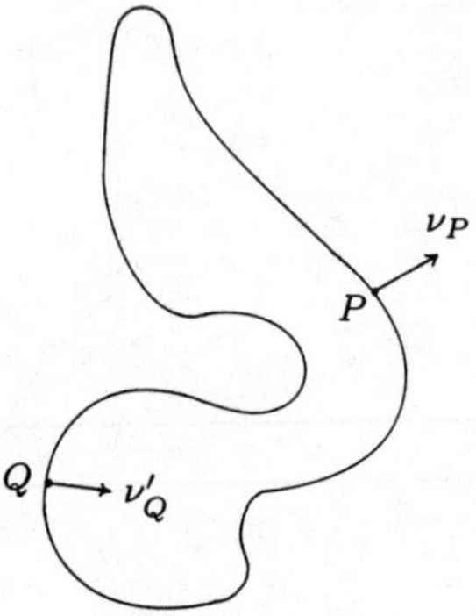

FIGURE 4

The functions

$$\partial U \ni P \longrightarrow \nu_P \in \mathbf{R}^2$$

and

$$\partial U \ni P \longrightarrow \nu'_P \in \mathbf{R}^2$$

are then continuously differentiable.

We shall need two analytic results about domains with C^2 boundary. They are formulated as Propositions 3 and 4.

□ **Proposition 3.** *If U is a domain with C^2 boundary then there is an open neighborhood W of ∂U such that if $z \in U \cap W$ then there is a unique point $P = P(z) \in \partial U$ which is nearest (in the sense of Euclidean distance) to z. In other words,*

$$\inf\{|z-Q| : Q \in \partial U\} = |z-P|.$$

We call W a tubular neighborhood of ∂U (see [MUN] for more on these matters).

PROOF.
Define

$$T : \partial U \times (-1,1) \longrightarrow \mathbf{C}$$
$$(Q,t) \longmapsto Q + t\nu_Q.$$

Think of T as a map from the two-dimensional real space $\partial U \times (-1,1)$ to the two-dimensional real space $\mathbf{C} \approx \mathbf{R}^2$. See Figure 5.

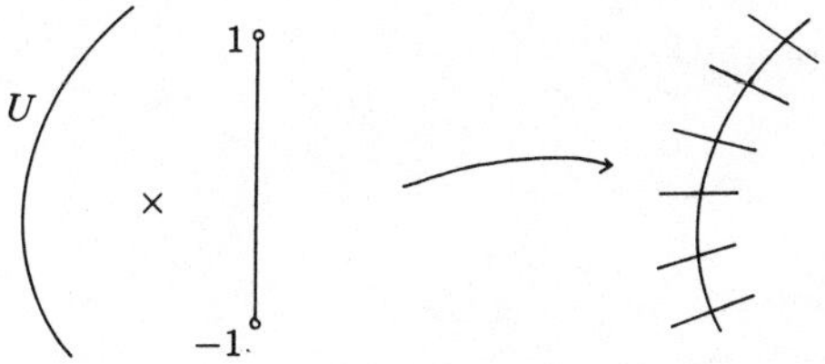

FIGURE 5

Fix a point $Q_0 \in \partial U$. Assume without loss of generality that the tangent line to ∂U at Q_0 is horizontal (this may be achieved by rotating the coordinate system). Then the Jacobian matrix of T at $(Q_0,0)$ is

$$\begin{pmatrix} 1 & 0 \\ 0 & 1 \end{pmatrix}$$

which is invertible. By the inverse function theorem (see [RU1] or [MUN]), there is a neighborhood V_0 of $(Q_0,0)$ in $\partial U \times (-1,1)$ on which T is invertible. Then $T(V_0) \equiv W(Q_0)$ is a neighborhood of Q_0 with the property that if $z \in W(Q_0)$, then there is a unique point $Q \in \partial U$ and a unique $t \in (-1,1)$ such that

$$z = Q + t\nu_Q.$$

Then Q is the unique nearest point in ∂U to z and the distance of z to Q is $|t|$. We set

$$W = \cup_{Q_0 \in \partial U} W(Q_0)$$

and we are finished. □

Remark.

The hypothesis of C^2 boundary was needed in order to apply the inverse function theorem. Explain why. □

□ **Proposition 4.** *Let $U \subseteq \mathbf{C}$ be a bounded domain with C^2 boundary. Then there is an $r_0 > 0$ such that for each $P \in \partial U$ there is a disc $D(C(P), r_0)$ of radius r_0 which is externally tangent to ∂U at P. There is also a disc $D(C'(P), r_0)$ which is internally tangent to ∂U at P. These discs have the further property that $\bar{D}(C(P), r_0) \cap \partial U = \{P\}$ and $\bar{D}(C'(P), r_0) \cap \partial U = \{P\}$.*

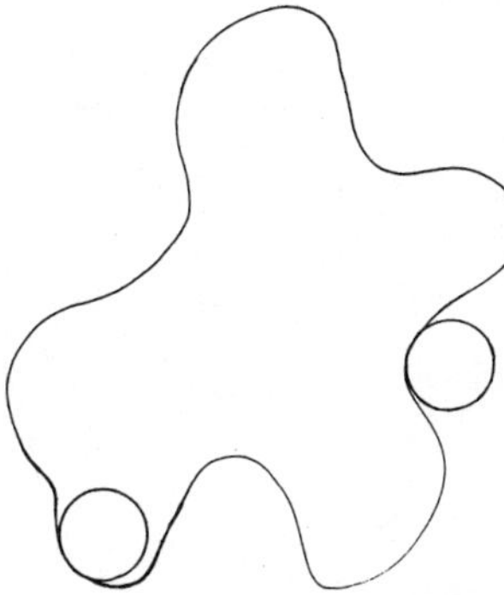

FIGURE 6

Remark.

Refer to Figure 6 and notice that a disc $\mathbf{d}$ is internally tangent to ∂U at P if $\mathbf{d} \subseteq U, \bar{\mathbf{d}} \ni P$, and $\partial\mathbf{d}$ and ∂U have the same tangent line at P.

A similar remark applies to externally tangent discs.

□

PROOF OF THE PROPOSITION.
This follows from basic calculus: To each point $P \in \partial U$ there corresponds a radius of curvature $r(P)$ and a center of curvature $c(P)$ (here we are using the word "curvature" in the classical Euclidean sense). The point $c(P)$ is $r(P)$ units along the principal normal from P.

The corresponding circle is called the circle of curvature. See [THO] for details. If the circle of curvature is *inside* the domain, take $C'(P) = c(P)$ and take $C(P)$ to be the reflected point in ∂U. See Figure 7. If the circle of

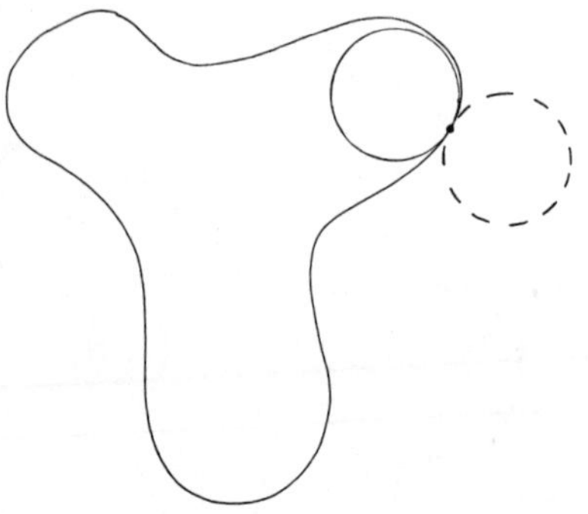

FIGURE 7

curvature is *outside* the domain, take $C(P) = c(P)$ and take $C'(P)$ to be a reflected point in ∂U.

Now, referring to [THO], we see that $r(P)$ depends on the second derivative of the boundary curve γ. In particular $r(P)$ is a continuous function of P. We take r_0 to be the (positive) minimum of r. We would like to think that we are finished.

Unfortunately, in spite of our best intentions, we may have a situation as in Figure 8 or 9: the circle of curvature has the right behavior *at* P, but it doesn't take into account the global behavior of U. As a result, it may be neither completely internal to U nor completely external to U. We therefore make the following adjustment. Let W be a tubular neighborhood of ∂U and choose $\epsilon > 0$ such that if $z \in \mathbf{C}$ has Euclidean distance less than ϵ from ∂U, then z is in W.

Define

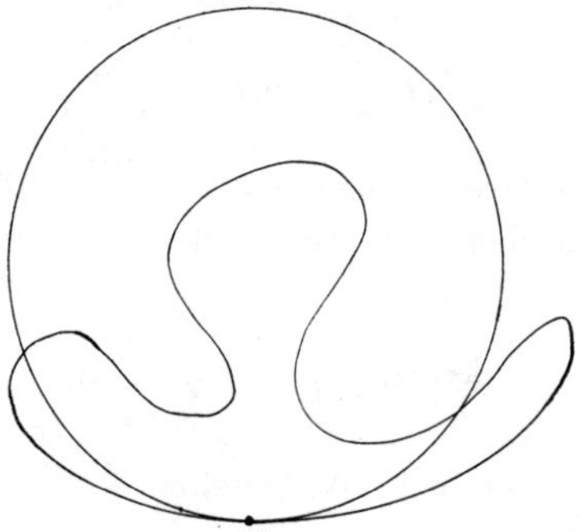

FIGURE 8

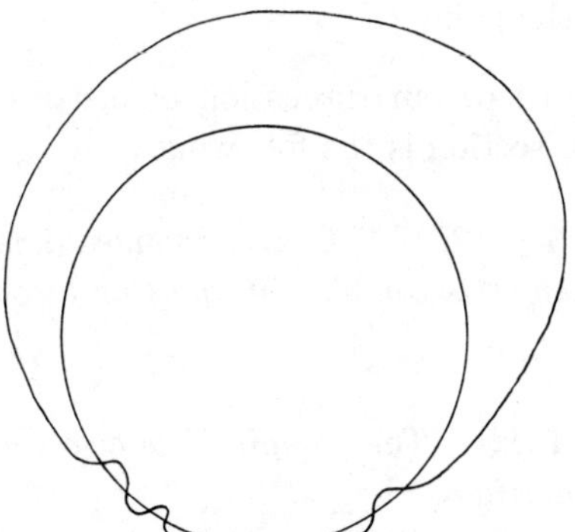

FIGURE 9

$$r^*(P) = \min\{\epsilon/2, r(P)\}.$$

This is a positive, continuous function of $P \in \partial U$. Hence there is a number $r_0 > 0$ such that

$$r^*(P) > r_0 \qquad \forall P \in \partial U.$$

This is the r_0 we seek. We redefine

$$C(P) = P + r_0\nu_P \quad \text{and} \quad C'(P) = P + r_0\nu'_P.$$

The definition of tubular neighborhood guarantees that a disc with center $C(P)$ or $C'(P)$ and radius r_0 can intersect ∂U only at the point P. □

We return to our discussion of metrics. The main result of this section is the following.

□ **Theorem 5.** *If $U \subseteq \mathbf{C}$ is a bounded domain with C^2 boundary then U is complete in the Carathéodory metric.*

□ **Corollary 5.1.** *The domain U is also complete in the Kobayashi metric.*

Proof of the Theorem.
Let $z \in U$ be in a tubular neighborhood W of ∂U, as provided by Proposition 3. Let P be the nearest boundary point to z and $D(C(P), r_0)$ the externally tangent disc provided by Proposition 4. The map

$$\begin{aligned}\mathbf{i}_P : U &\longrightarrow D(C(P),r_0)\\ \zeta &\longmapsto C(P) + \frac{(r_0)^2}{\zeta - C(P)}\end{aligned}$$

inverts U into the disc $D(C(P),r_0)$. The map

$$\begin{aligned}\mathbf{j}_P : D(C(P),r_0) &\longrightarrow D(0,1)\\ \zeta &\longmapsto \frac{\zeta - C(P)}{r_0}\end{aligned}$$

is holomorphic. To estimate the Carathéodory metric at z, we use the distance decreasing property of the Carathéodory metric:

$$\begin{aligned}F_U^C(z) &\geq \left((\mathbf{j}_P \circ \mathbf{i}_P)^* F_C^{D(0,1)}\right)(z)\\ &\equiv |(\mathbf{j}_P \circ \mathbf{i}_P)'(z)| F_C^{D(0,1)}(\mathbf{j}_P(\mathbf{i}_P(z))).\end{aligned} \qquad (*)$$

We now estimate the various terms in this last expression.

First,

$$\begin{aligned}(\mathbf{j}_P \circ \mathbf{i}_P)'(z) &= \frac{1}{r_0} \cdot (\mathbf{i}_P)'(z)\\ &= \frac{1}{r_0} \cdot \frac{-(r_0)^2}{(z - C(P))^2}\\ &= \frac{-r_0}{(z - C(P))^2}\\ &= \frac{-r_0}{(\delta + r_0)^2},\end{aligned} \qquad (**)$$

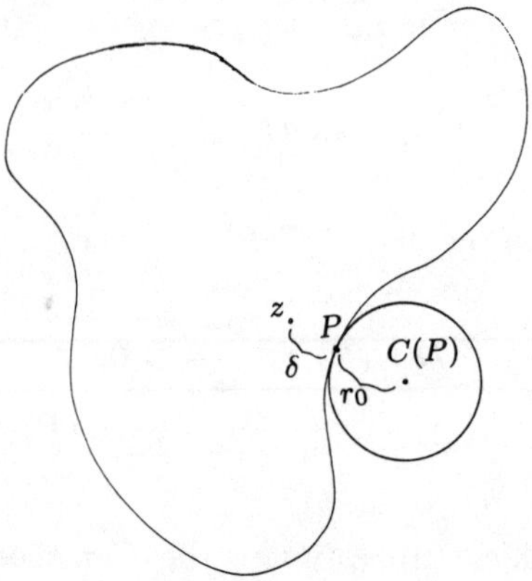

FIGURE 10

where δ is the Euclidean distance of z to P. See Figure 10. Also

$$\mathbf{j}_P \circ \mathbf{i}_P(z) = \mathbf{j}_P\left(\frac{(r_0)^2}{z - C(P)} + C(P)\right)$$
$$= \frac{r_0}{z - C(P)}$$

hence

$$|\mathbf{j}_P \circ \mathbf{i}_P(z)| = \frac{r_0}{\delta + r_0} = 1 - \frac{\delta}{r_0 + \delta}.$$

Recall that, on the unit disc, the Carathéodory metric coincides with the Poincaré metric. It follows from our calculation of the Poincaré metric in Proposition 2 of Section 1.3 that

$$\begin{aligned} F_C^{D(0,1)}(\mathbf{j}_P \circ \mathbf{i}_P(z)) &= \frac{1}{(\delta/(r_0+\delta))(2-\delta/(r_0+\delta))} \\ &\geq \frac{r_0+\delta}{2\delta} \qquad (***) \\ &\geq \frac{r_0}{2} \cdot \frac{1}{\delta}. \end{aligned}$$

In summary, using $(*), (**)$, and $(***)$, we have

$$F_C^U(z) \geq \frac{r_0}{(\delta+r_0)^2} \cdot \frac{r_0}{2} \cdot \frac{1}{\delta} \geq C_0 \cdot \frac{1}{\delta},$$

where C_0 is a positive constant that depends only on r_0.

But this is precisely the estimate which enabled us, in Section 1.4, to prove that the disc is complete in the Poincaré metric. We leave it now as an exercise to provide the details which show that the Carathéodory distance from any fixed interior point $P_0 \in U$ to a point with Euclidean distance δ from the boundary has size $C \cdot \log(1/\delta)$ and to conclude that U is complete in the Carathéodory metric. □

Exercise.

Exploit the internally tangent disc at each boundary point to prove that there is a constant C_1 such that

$$F_C^U(z) \leq C_1 \cdot \frac{1}{\delta}.$$

(Hint: Use the inclusion map from the internal disc to the domain, together with the distance decreasing property of the Carathéodory metric.) □

Exercise.
Let

$$U = D \setminus \{0\},$$

the punctured disc.

FIGURE 11

See Figure 11. This domain does *not* have C^2 boundary. Use the Riemann removable singularities theorem and Cauchy estimates to determine the behavior of the

Carathéodory metric near the boundary point 0 of U. Conclude that U is not complete in the Carathéodory metric. What can you say about the Kobayashi metric? □

□ **Corollary to Theorem 5.** *Let U be any finitely connected region in $\mathbf{C}$ (that is, the complement of U has finitely many components). Further assume that each component of $\hat{\mathbf{C}} \setminus U$ is a continuum. Then the Carathéodory metric on U is complete.*

PROOF.
Since completeness of the Carathéodory metric is a conformal invariant, it is enough to show that U is conformally equivalent to a domain U' having the property that ∂U has C^2 boundary. If ∂U has k components then k successive applications of the Riemann mapping theorem give rise to a conformal mapping of U to a region comprised of the unit disc with $k-1$ pairwise disjoint subdiscs removed. That completes the proof. □

We now use the completeness of the Carathéodory metric to prove a version of the Lindelöf principle using the geometric point of view. First we need some terminology.

Let U be a domain with C^2 boundary. Choose a point $P \in \partial U$ and let ν_P be the unit outward normal. If f is a continuous function on U we say that f has *radial boundary limit ℓ* at P provided that

$$\lim_{r \to 0^+} f(P - r\nu_P) = \ell.$$

If $\alpha > 1$, set

$$\Gamma_\alpha(P) = \{z \in U : |z - P| < \alpha \cdot \operatorname{dist}(z, \partial U)\}.$$

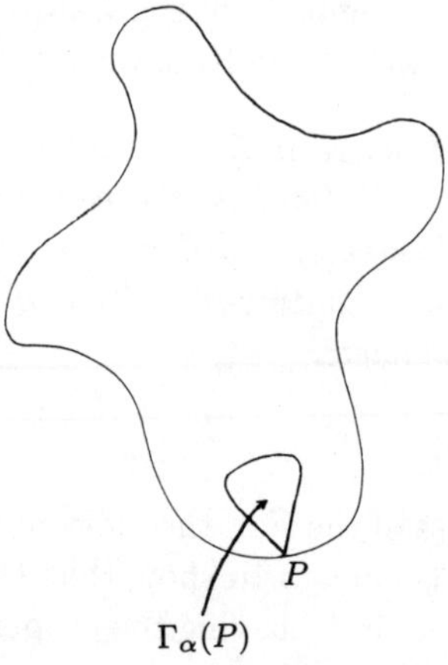

FIGURE 12

Here the expression $\operatorname{dist}(z, \partial U)$ denotes Euclidean distance of z to ∂U. We call the domain Γ_α a *Stolz region* or *nontangential* approach region at the point P. See Figure 12. We say that the function f has *nontangential limit* ℓ at P if for each $\alpha > 1$ it holds that

$$\lim_{\Gamma_\alpha(P) \ni z \to P} f(z) = \ell.$$

Clearly the possession of a nontangential limit at P is a stronger condition than the possession of a radial limit at P. For example, on the unit disc the continuous function

$$f(z) = \frac{y}{1-x}$$

has radial limit 0 at the boundary point $P = 1+i0$. However it does not possess a nontangential limit at P (exercise—try $P_j = (1-1/j)+i/j$ for $j = 1,2,\ldots$). It is therefore surprising that the Lindelöf principle tells us that for bounded holomorphic functions the two notions of boundary limit coincide:

□ **Theorem 6.** *Let U_1, U_2 be bounded domains with C^2 boundary and let*

$$f : U_1 \longrightarrow U_2$$

be holomorphic. If $P \in \partial U_1$, $Q \in \partial U_2$, and f has radial limit Q at P then f has nontangential limit Q at P.

PROOF.
If z is an element of one of our domains U_j and if $s > 0$ then we let $B(z,s)$ denote the metric ball with center z and radius s in the Carathéodory metric for U_j. For $P \in \partial U_1$ and $\beta > 0$ fixed, we define

$$\mathcal{M}_\beta(P) = \cup_{0<r<1} B(P - r\nu_P, \beta).$$

The estimate

$$F_C^U(z) \approx \frac{C}{\operatorname{dist}(z, \partial U)} \qquad (*)$$

makes it a simple matter to calculate that the regions $\mathcal{M}_\beta$ are comparable to the regions Γ_α (in this last formula, and in what follows, "dist" means Euclidean distance). That is,

$$\lim_{\Gamma_\alpha(P)\ni z\in P} f(z) = \ell, \qquad \forall \alpha > 1$$

iff

$$\lim_{\mathcal{M}_\beta(P)\ni z\to P} f(z) = \ell, \qquad \forall \beta > 0. \qquad (**)$$

Thus it is enough to prove $(**)$.

Since

$$\mathcal{M}_\beta(P) = \cup_{0<r<1} B(P - r\nu_P, \beta),$$

the distance decreasing property of f with respect to the Carathéodory metric implies that

$$f(\mathcal{M}_\beta(P)) \subseteq \cup_{0<r<1} B(f(P - r\nu_P), \beta).$$

Pick $\epsilon > 0$. By the radial limit hypothesis, there is a $\delta > 0$ such that if $0 < t < \delta$ then

$$|f(P - t\nu_P) - Q| < \epsilon.$$

For such a t, if $z \in B(P - t\nu_P, \beta)$ then

$$f(z) \in B(f(P - t\nu_P), \beta).$$

But

$$\mathrm{dist}(f(P - t\nu_P), \partial U_2) \le \mathrm{dist}(f(P - t\nu_P), Q) < \epsilon.$$

Therefore the estimate $(*)$ implies that the metric ball $B(f(P-t\nu_P),\beta)$ has Euclidean radius not exceeding $C\cdot\epsilon$. Here C depends on β, but β has been fixed once and for all. Thus

$$|f(z)-f(P-t\nu_P)|<C\epsilon, \qquad \forall z\in B(P-t\nu_P,\beta).$$

We conclude that

$$\begin{aligned}|f(z)-Q| &\le |f(z)-f(P-t\nu_P)|+|f(P-t\nu_P)-Q| \\ &\le C\epsilon+\epsilon=C'\epsilon.\end{aligned}$$

This is the desired conclusion. □

It is not difficult to see that the nontangential approach is the broadest possible method for calculating boundary limits of bounded holomorphic functions. See [PRI] for explicit constructions of examples to show that the nontangential approach is the sharp condition. This reference also contains a complete account of the positive results, as well as the history, of the theory of boundary behavior of holomorphic functions.

4. An Application of Completeness: the Automorphism Group of a Domain

This section is an introduction to a topic which is usually treated as part of Riemann surface theory. (See

[FK] for a nice treatment of the Riemann surface point of view.) We shall not use any of this theory, nor shall we make any reference to Riemann surfaces. Instead, we use geometry.

Definition 1.

Let $U \subseteq \mathbf{C}$ be a domain. We let $\mathrm{Aut}(U)$ denote the family of conformal self-maps of U (i.e., one-to-one, onto holomorphic functions from U to U).

□ **Proposition 2.** *With the binary operation being the composition of functions, the set* $\mathrm{Aut}(U)$ *forms a group. We call this group the automorphism group of* U.

PROOF.
First, the identity function $id : U \to U$, $id(z) = z$, is the group identity. Now if $\phi \in \mathrm{Aut}(U)$ then ϕ is one-to-one and onto so that the inverse function ϕ^{-1} makes sense and is holomorphic; it is also one-to-one and onto. And $\phi \circ \phi^{-1} = \phi^{-1} \circ \phi = id$. Thus the inverse function ϕ^{-1} is the group inverse.

Next, let $\phi, \psi \in \mathrm{Aut}(U)$. Then, by definition, $\phi : U \to U$ and $\psi : U \to U$ so certainly $\psi \circ \phi : U \to U$. Since ϕ and ψ are one-to-one, then so is the composition $\psi \circ \phi$. Likewise, the composition is onto because each of the individual functions is. Thus $\mathrm{Aut}(U)$ is closed under the group operation.

Finally, the group operation is associative because composition of functions is associative. Hence $\mathrm{Aut}(U)$ is a group. □

Example 1.

Let U be the disc. By the exercise following Theorem 3 of Section 0.2, $\mathrm{Aut}(U)$ consists of all maps of the form

$$\zeta \longmapsto \mu \cdot \frac{\zeta - a}{1 - \bar{a}\zeta}$$

where $a, \mu \in \mathbf{C}$, $|a| < 1$, and $|\mu| = 1$. □

Example 2.

Let $0 < r < R < \infty$ and let

$$A_{r,R} = \{z : r < |z| < R\}.$$

Then $\mathrm{Aut}(A_{r,R})$ consists only of the rotations

$$z \longmapsto \mu \cdot z \quad (\text{for } \mu \text{ a unimodular constant}),$$

the reflection

$$\sigma : z \longmapsto \frac{R \cdot r}{z},$$

and compositions of these two types of functions.

Clearly each of these two types of functions are elements of $\mathrm{Aut}(A_{r,R})$. For the converse, let $\phi \in \mathrm{Aut}(A_{r,R})$. Notice that the Schwarz reflection principle implies that ϕ continues analytically past $\partial A_{r,R}$. The maximum/minimum principle then guarantees that $\partial A_{r,R}$ gets mapped by ϕ to $\partial A_{r,R}$. After composing with the inversion σ, we may suppose that $\{z : |z| = r\}$ is mapped to itself and that $\{z : |z| = R\}$ is mapped to itself. We claim that ϕ must be a rotation.

We may reflect ϕ across both boundaries of $A_{r,R}$ to obtain an automorphism of the larger annulus $\{z : r^2/R < |z| < R^2/r\}$. Continuing this reflection process countably many times, we can analytically continue ϕ to an automorphism of $\mathbf{C}' \equiv \mathbf{C} \setminus \{0\}$. By the Riemann removable singularities theorem, ϕ continues analytically to an automorphism of $\mathbf{C}$ which fixes the origin. Thus there is a constant α such that $\phi(z) \equiv \alpha z$. Since ϕ is the continuation of an automorphism of a proper annulus $A_{r,R}$, it must be that α has modulus 1. Therefore ϕ is a rotation as claimed.

□

Given a domain U, we equip $\text{Aut}(U)$ with the topology of uniform convergence on compact sets. That is,

$$\text{Aut}(U) \ni \phi_j \longrightarrow \phi \in \text{Aut}(U)$$

provided that $\phi_j \to \phi$ normally. We are particularly interested in the question of when $\text{Aut}(U)$ is compact. In the given topology, $\text{Aut}(U)$ is compact if and only if every sequence $\{\phi_j\} \subseteq \text{Aut}(U)$ has a subsequence which converges normally *to an element of* $\text{Aut}(U)$.

Example 3.

Let $0 < r < R < \infty$ and let

$$A_{r,R} = \{z : r < |z| < R\}$$

be the corresponding annulus. Then $\text{Aut}(A_{r,R})$ is compact. To see this, let α_j be a sequence of automorphisms of $A_{r,R}$.

Now each of the α_j is either a rotation, the reflection σ, or a composition of the two. Thus infinitely many of the α_j are of the same one of these three types. If infinitely many are equal to the reflection σ, then the subsequence consisting of these elements (this will be a constant sequence) converges to σ. If infinitely many consist of a rotation, then let us write these as

$$\alpha_{j_k}(z) = \mu_{j_k} \cdot z,$$

where the μ_{j_k} are unimodular constants. Since the unit circle is compact, there is a subsequence—call it μ_m—which converges to a unimodular constant μ_0. But then the corresponding automorphisms

$$\alpha_m(z) = \mu_m z$$

converge normally to the limit automorphism $\alpha_0(z) = \mu_0 z$. The third possibility is that infinitely many of the α_j consist of a rotation followed by the reflection σ. Write these as

$$\alpha_{j_k}(z) = \sigma(\mu_{j_k} z),$$

with the μ_{j_k} being unimodular constants. We extract a convergent subsequence $\{\mu_m\}$ with limit μ_0 and define

$$\alpha_0(z) = \sigma(\mu_0 z).$$

Then the subsequence

$$\alpha_m(z) = \sigma(\mu_m z)$$

converges normally to $\alpha_0(z)$.

The three possibilities having been covered, we conclude that $\mathrm{Aut}(A_{r,R})$ is compact. □

Remark.

It is an interesting fact that if the connectivity of a planar domain is finite but greater than two then the automorphism group of the domain is a finite group (see [FK]). It is an open problem to determine *which* finite groups arise as automorphism groups of planar domains. □

Example 4.

Let $U = D$, the unit disc. Then the functions

$$\beta_j(\zeta) = \frac{\zeta + (1 - 1/j)}{1 + (1 - 1/j)\zeta}$$

are elements of $\mathrm{Aut}(D)$. However $\{\beta_j\}$ converges normally to the function

$$\beta_0(\zeta) \equiv 1.$$

Notice that $\beta_0 \notin \mathrm{Aut}(D)$—in fact it is a constant function. Therefore $\mathrm{Aut}(D)$ is not compact. □

Remark.

Notice that, at least when U is bounded, the issue of whether $\mathrm{Aut}(U)$ is compact is not a question of whether $\{\phi_j\} \subseteq \mathrm{Aut}(U)$ has a subsequence that converges: by normal families this is essentially automatic. Rather, the issue is *whether the subsequence will converge to an element of* $\mathrm{Aut}(U)$. It is this last phenomenon that fails for the disc, but works for the annulus. □

The principal result of this section is that the disc is characterized by the noncompactness of its automorphism group. More precisely, we have

□ **Theorem 3.** *Let $U \subseteq \mathbf{C}$ be a bounded domain with C^2 boundary. If $\mathrm{Aut}(U)$ is noncompact then U is conformally equivalent to the unit disc.*

We prove this theorem with a sequence of lemmas, each of which has intrinsic interest.

□ **Lemma 4.** *Let $U \subseteq \mathbf{C}$ be bounded. The group $\mathrm{Aut}(U)$ is compact if and only if for each $P \in U$ there is a compact $K_P \in U$ such that $\phi(P) \in K_P$ for all $\phi \in \mathrm{Aut}(U)$.*

PROOF.

Assume that $\mathrm{Aut}(U)$ is compact. Fix $P \in U$. If there is no set K_P as claimed then there exist $\phi_j \in \mathrm{Aut}(U)$ such that $\phi_j(P) \to w \in \partial U$, some w. But U is bounded so that $\{\phi_j\}$ is a normal family; thus there is a subsequence ϕ_{j_k} and a limit function ϕ_0 such that

$$\phi_{j_k} \longrightarrow \phi_0$$

normally.

Notice that the image of each ϕ_j lies in U hence the image of ϕ_0 lies in the *closure* $\bar{U}$ of U. If ϕ_0 is nonconstant then it satisfies the open mapping principle. But

$$\phi_0(P) = \lim_{k\to\infty} \phi_{j_k}(P) = w,$$

hence the image of ϕ_0 contains the accumulation point $w \in \partial U$, so it contains a neighborhood of w. This is impossible because w is in the boundary of the image of ϕ_0. Therefore ϕ_0 must be constantly equal to w; thus $\phi_0 \notin \mathrm{Aut}(U)$. The sequence ϕ_{j_k} therefore violates the compactness of $\mathrm{Aut}(U)$. We conclude that K_P must exist.

For the converse, fix $P \in U$ and let K_P be the corresponding compact set in U whose existence we assume. Let $\{\phi_j\} \subseteq \mathrm{Aut}(U)$ be any sequence. Since U is bounded, there is a normally converging subsequence ϕ_{j_k} with limit function ϕ_0. As in the first half of the proof, if the image of ϕ_0 contains any boundary point w then ϕ_0 must be constantly equal to w. But the image of P under ϕ_0 must lie in K_P, so this possibility is ruled out. We conclude that the image of ϕ_0 lies in U.

Next notice that each ϕ_{j_k} has an inverse ψ_{j_k}. Passing to another subsequence, we may suppose that the ψ_{j_k} converge to a limit function ψ_0. For convenience, we denote this final subsequence by ψ_m, corresponding to the functions ϕ_m. Just as for ϕ_0, we can be sure that the image of ψ_0 lies in the interior of U.

Now we have

$$z \equiv \lim_{m\to\infty} \phi_m \circ \psi_m(z) = \phi_0 \circ \psi_0(z).$$

Since $\mathbf{i}(z) \equiv z$ is onto, so is ϕ_0. Also, as in the proof of Theorem 5 in Section 2, the image of ψ_0 is open, closed, and nonempty. Therefore, ψ_0 is surjective. Since $\mathbf{i}(z)$ is injective, it now follows that ϕ_0 is injective. Therefore, $\phi_0 \in \mathrm{Aut}(U)$. We conclude that

$$\mathrm{Aut}(U) \ni \phi_m \longrightarrow \phi_0 \in \mathrm{Aut}(U),$$

and $\mathrm{Aut}(U)$ is compact. □

□ **Lemma 5.** *Let $U \subseteq \mathbf{C}$ be a bounded domain with C^2 boundary. Suppose that $P \in U$, $\{\phi_j\}$ are holomorphic maps from U to U, and*

$$\phi_j(P) \longrightarrow w \in \partial U.$$

If K is compact in U and V is a neighborhood of w then there exists a positive number J such that if $j \geq J$ then

$$\phi_j(K) \subseteq V.$$

See Figure 1.

PROOF.
Since U, when equipped with the Carathéodory metric, is a metric space and since K is compact, there is a positive number R such that the metric ball $B(P,R)$ contains K. Let $Q_j = \phi_j(P)$. Since each ϕ_j is distance decreasing in the Carathéodory metric, it follows that $\phi_j(B(P,R)) \subseteq B(Q_j,R)$. We claim that there is a positive J such that whenever $j \geq J$ then $B(Q_j,R) \subseteq V$. Assuming the claim, we would then have

$$\phi_j(K) \subseteq \phi_j(B(P,R)) \subseteq B(Q_j,R) \subseteq V,$$

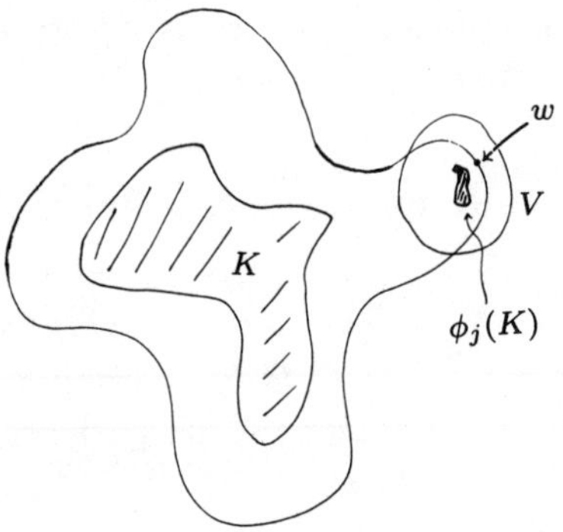

FIGURE 1

as required.

To prove the claim, recall from the proof of Theorem 5 in the last section that the Euclidean radii of the metric balls $B(Q_j, R)$ must tend to 0. Choose $\epsilon > 0$ such that the Euclidean disc of center w and radius 2ϵ lies in V. We select J so large that when $j > J$ then both the Euclidean distance of Q_j to w is less than ϵ and the Euclidean radius of $B(Q_j, R)$ is less than ϵ. The claim now follows from the triangle inequality. □

PROOF OF THEOREM 3.
If $\operatorname{Aut} U$ is not compact then, by Lemma 4, there is a sequence $\phi_j \in \operatorname{Aut}(U)$ and a $P \in U$ such that

$$\phi_j(P) \longrightarrow w \in \partial U,$$

for some $w \in \partial U$.

Let

$$\gamma : [0,1] \longrightarrow U$$

be any continuous closed curve in U. Since ∂U is C^2 there is a neighborhood V of w such that $U \cap V$ is simply connected (see Figure 2—the existence of the interior osculating circle provided by Proposition 3 of Section 3 makes this assertion clear).

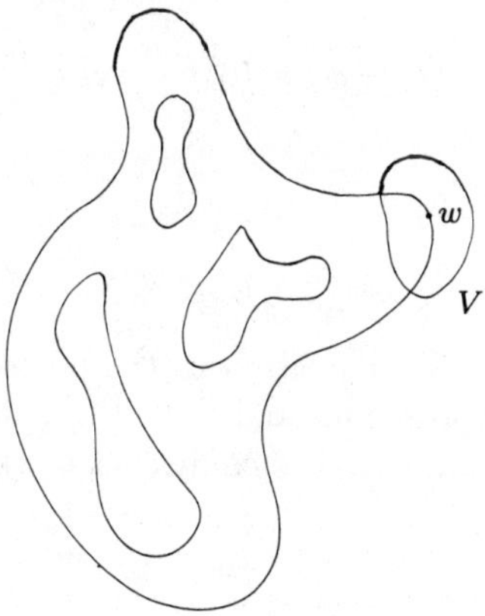

FIGURE 2

Let

$$K = \{\gamma(t) : 0 \leq t \leq 1\}.$$

Then K is compact. By Lemma 5, there is a $J \geq 0$ such that $j \geq J$ implies $\phi_j(K) \subseteq U \cap V$. Thus $\phi_j \circ \gamma$ is a continuous, closed curve in $U \cap V$. The simple connectivity of

$U \cap V$ implies that $\phi_j \circ \gamma$ may be continuously deformed to the point $\phi_j \circ \gamma(0)$; that is, there is a homotopy

$$\Psi : [0,1] \times [0,1] \longrightarrow U \cap V$$

such that

$$\Psi(0,t) = \phi_j \circ \gamma(t), \qquad \forall t \in [0,1]$$

and

$$\Psi(1,t) = \phi_j \circ \gamma(0), \qquad \forall t \in [0,1].$$

But then

$$(\phi_j)^{-1} \circ \Psi$$

is a homotopy of the curve γ to the point $\gamma(0)$. It follows that U is simply connected. By the Riemann mapping theorem, U is conformally equivalent to the disc. □

Remark.

It is possible to prove this theorem without resort to the Riemann mapping theorem. Instead one uses a delicate approximation argument on a decreasing sequence of boundary neighborhoods $V_j \cap U$ of w to prove that

$$F_C^U(P) = F_K^U(P).$$

It then follows from Theorem 5 of Section 2 that U is conformally equivalent to the disc. In the same vein, one can actually prove the Riemann mapping theorem using F_C^U and F_K^U. Since the ideas involved are rather complicated and would take us far afield, we omit them. □

The notion of compact automorphism group is a bit abstract, so we conclude this section with an interpretation of Theorem 3 which is more concrete. Let us say that $\mathrm{Aut}(U)$ *acts transitively* on the domain U if for any P, Q in U there is an element $\phi \in \mathrm{Aut}(U)$ such that $\phi(P) = Q$. In other words, the automorphism group acts transitively if it is sufficiently rich with elements that it can move any point of U to any other.

Example 5.

Let $U = D$, the unit disc. If P is any element of D then the automorphism

$$\phi_P(\zeta) \equiv \frac{\zeta - P}{1 - \bar{P}\zeta}$$

has the property that $\phi_P(P) = 0$. Now if P and Q are arbitrary elements of D then $(\phi_Q)^{-1} \circ \phi_P$ is an automorphism of D which maps P to Q. Thus the automorphism group of the disc *does* act transitively. □

Example 6.

Let $A_{r,R}$ be the annulus

$$\{z : r < |z| < R\}.$$

Then $\mathrm{Aut}(A_{r,R})$ does *not* act transitively on $A_{r,R}$. In fact if $P,Q \in A_{r,R}$, if $|P| \neq |Q|$, and if $|P| \neq r \cdot R/|Q|$ then Example 2 shows that there is no automorphism of $A_{r,R}$ which takes P to Q. □

We can now give a consequence of Theorem 3, using the language of transitivity.

□ **Corollary 6.** *Let U be a bounded domain with C^2 boundary. If $\mathrm{Aut}(U)$ acts transitively on U then U is conformally equivalent with the disc.*

PROOF.
Fix a point P in U. Since $\mathrm{Aut}(U)$ acts transitively on U, P can be moved to any other point of U by some element of $\mathrm{Aut}(U)$. Thus $\{\phi(P) : \phi \in \mathrm{Aut}(U)\}$ can be contained in no compact set $K_P \subseteq U$. By Lemma 4, $\mathrm{Aut}(U)$ is noncompact. By the theorem, U must be conformally equivalent to the disc. □

It is natural to wonder whether the hypothesis of the C^2 boundary is really necessary for the results presented in this section. The answer is that *some* regularity of the boundary is required, as the next example illustrates.

Example 7.
Define the mapping

$$\phi(z) = \frac{z + 1/2}{1 + (1/2)z}.$$

We know that ϕ is an automorphism of the disc. Define $\phi^2 = \phi \circ \phi$, $\phi^3 = \phi \circ \phi \circ \phi$, etc. With ϕ^{-1} denoting the inverse of ϕ, we set $\phi^{-2} = \phi^{-1} \circ \phi^{-1}$, $\phi^{-3} = \phi^{-1} \circ \phi^{-1} \circ \phi^{-1}$, etc. We let $\phi^0(z) \equiv z$.

For $j \in \mathbf{Z}$ we define

$$D_j = \{\phi^j(z) : |z| \leq 1/4\}.$$

Finally, let

$$U = D(0,1) \setminus \cup_{j \in \mathbf{Z}} D_j.$$

The domain U is displayed in Figure 3.

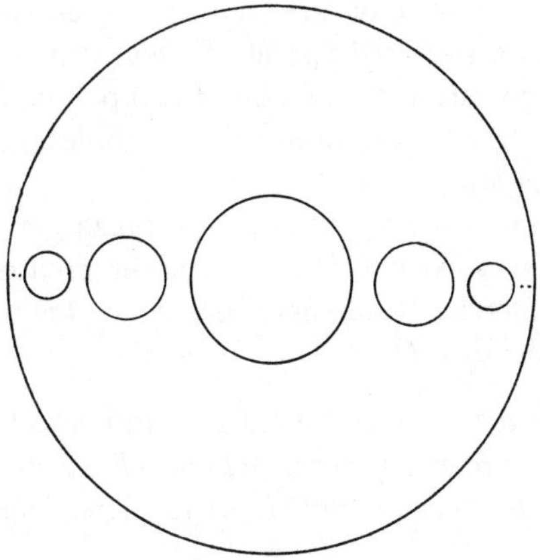

FIGURE 3

It is not difficult to verify that $\mathrm{Aut}(U)$ consists precisely of the maps $\phi^j, j \in \mathbf{Z}$. Thus $\mathrm{Aut}(U)$ is noncompact (exercise). Yet U is certainly not conformally equivalent to the disc.

Notice that the boundary of U is not C^2, so there is no contradiction with Theorem 3. □

5. Hyperbolicity and Curvature

In this section we relate the concepts of curvature and normal families to the property of nondegeneracy of the Kobayashi metric. A consequence is that we can prove that the plane and $\mathbf{C} \setminus \{0\}$ cannot support a metric of strictly negative curvature. Thus we see, from a geometric point of view, why Picard's theorem must specify the omission of *two* points of the range of an entire function in order for that function to be necessarily constant.

We begin with an elegant generalization of the fact that the Carathéodory metric is never greater than the Kobayashi metric. Recall here that ρ denotes the Poincaré metric on the disc D.

□ **Proposition 1.** *If $U \subseteq \mathbf{C}$ is a domain which is equipped with a metric σ and if every holomorphic map $f : D \to U$ is distance decreasing from (D,ρ) to (U,σ) then*

$$\sigma \leq F_K^U.$$

PROOF.
Abbreviate F_K^U by the symbol μ. Let $z \in U$. If $\phi : D \to U$ and $\phi(0) = z$ then, by hypothesis,

$$\phi^*\sigma(0) \le \rho(0)$$

or

$$\frac{1}{|\phi'(0)|} \ge \frac{\sigma(z)}{\rho(0)} = \sigma(z).$$

Taking the infimum over all ϕ we conclude, for any $\xi \in \mathbf{C}$, that

$$\|\xi\|_{\mu,z} = \inf_{\phi \in (U,D)_z} \frac{|\xi|}{|\phi'(0)|} \ge |\xi| \cdot \sigma(z) = \|\xi\|_{\sigma,z},$$

as desired. □

Definition 2.

A domain $U \subseteq \mathbf{C}$ is called *hyperbolic* if the Kobayashi distance on U is actually a distance, that is, if

$$d_{\mathrm{Kob}}(P,Q) > 0$$

whenever P and Q are distinct points of U.

□ **Proposition 3.** *Let $U \subseteq \mathbf{C}$ be a domain equipped with a metric σ which has the property that its curvature $\kappa = \kappa_\sigma$ satisfies*

$$\kappa \le -B < 0$$

for some positive constant B. Then U is hyperbolic.

PROOF.
Let f be any map of the disc D into U. By Ahlfors's version of Schwarz's lemma, we have that

$$f^*\sigma \leq \frac{\sqrt{4}}{\sqrt{B}}\rho,$$

where ρ is the Poincaré metric on the disc. Replace σ by $\tilde{\sigma} \equiv (\sqrt{B}/\sqrt{4}) \cdot \sigma$. Then our inequality is

$$f^*\tilde{\sigma} \leq \rho.$$

Thus f is distance decreasing from (D,ρ) to $(U,\tilde{\sigma})$. By Proposition 1, we may conclude that

$$\tilde{\sigma} \leq F_K^U$$

or

$$C \cdot \sigma \leq F_K^U,$$

where $C = \sqrt{B}/\sqrt{4}$ is a positive constant. Thus the Kobayashi distance is bounded from below by a positive constant times the σ-distance. Since σ is assumed to be nondegenerate, then so is F_K^U. □

□ **Corollary 3.1.** *Neither the plane* $\mathbf{C}$ *nor the punctured plane* $\mathbf{C} \setminus \{0\}$ *possesses a metric of strictly negative curvature.*

PROOF.
Neither $\mathbf{C}$ nor $\mathbf{C} \setminus \{0\}$ is hyperbolic. □

□ **Corollary 3.2.** *If* $U = \mathbf{C} \setminus \{P_1, \ldots, P_k\}$, *with the* P_j*'s distinct and* $k \geq 2$, *then* U *is hyperbolic.*

PROOF.
As we saw in Chapter 2, such a U has a metric of strictly negative curvature. □

Remark.

An alternate proof of Corollary 3.2 may be obtained by noticing that a domain U which satisfies the hypothesis will imbed into $\mathbf{C} \setminus \{0,1\}$ and then using the distance decreasing property of the Kobayashi metric. □

Now we will connect hyperbolicity with the concept of normal families. We first need some new terminology.

Definition 4.

A domain U is *taut* if the family $\mathcal{F}$ of holomorphic functions from the disc D into U is normal.

Example 1.

The disc D is taut. For if $\mathcal{F}$ is the family of holomorphic functions from D to D then the elements of $\mathcal{F}$ are uniformly bounded by 1. So there is a sequence of elements $\{f_j\} \subseteq \mathcal{F}$ which converges to a limit function f_0. The function f_0 is certainly holomorphic, and there are two possibilities: i) the image of f_0 contains a boundary element w, in which case the open mapping principle implies that f_0 is constant (this is the compactly divergent case); (ii) the image of f_0 lies entirely in the interior of D, in which case $f_0 \in \mathcal{F}$ (this is the convergent case). □

Example 2.

The domain $U = \mathbf{C} \setminus \{0\}$ is not taut. For the sequence of functions

$$f_n(z) = e^{nz}$$

maps the disc into U but there is neither a convergent subsequence nor a compactly divergent one (the origin is mapped to 1 by every f_n but $f_n(1/2) \to \infty$ as $n \to \infty$). □

□ **Proposition 5.** *A planar domain U is hyperbolic if and only if it is taut.*

PROOF.
We know that if U is hyperbolic then $\mathbf{C} \setminus U$ must contain at least two points. By the general version of Montel's theorem in Section 2.4, U must therefore be taut.

Conversely, assume that U is taut. Then, by Example 2, $\mathbf{C} \setminus U$ must contain at least two points. But then U is hyperbolic. □

Remark.

One theme that comes through in this section is that the only enemy is $\mathbf{C}$ or $\mathbf{C} \setminus \{0\}$. As soon as a domain excludes at least two points, then it has all relevant properties. In the theory of Riemann surfaces it also holds that a domain is taut if and only if it is hyperbolic (see [FK]). However in several complex variables matters are no longer so simple. In that context, there are infinitely many distinct domains which possess one of the properties (tautness, hyperbolicity) but not the other. □

Exercise.

The domain

$$U = \{z \in \mathbf{C} : |z| < 1\} \setminus \{x + i0 : 0 \leq x < 1\}$$

is taut and hyperbolic. □

CHAPTER 4

A GLIMPSE OF SEVERAL COMPLEX VARIABLES

0. Functions of Several Complex Variables

At a naive level, the analysis of several real variables is much like the real analysis of one variable; however, one deals with n-tuples of reals instead of scalars and one needs matrices to keep track of information. Of course deeper study reveals much complexity and richness in analysis of several real variables. It is noteworthy that, for the most part, this richness was discovered rather late in the history of the subject—mostly in the last forty years.

The history of analysis of several complex variables is quite different. Early in the subject, in the first decade of this century, two remarkable discoveries indicated that this area has an incredible depth and variety which one complex variable does not even hint at. Even today we have only scratched the surface of several complex variables.

Let us briefly discuss the two developments which established several complex variables as a subject in its own right. The first is related to the Riemann mapping theorem. As we have discussed throughout the present volume, the theorem asserts that, with the exception of the plane, any domain topologically equivalent to the disc is conformally equivalent to it. One might expect an analogue to this result in two complex variables. But, to begin with, what is the analogue of the disc in $\mathbf{C}^2$? Two candidates come to mind: the ball

$$\{(z_1, z_2) \in C^2 : |z_1|^2 + |z_1|^2 < 1\}$$

and the bidisc

$$\{(z_1, z_2) \in C^2 : |z_1| < 1,\ |z_2| < 1\}.$$

Which of these should serve as the model domain for the multivariable Riemann mapping theorem? Before answering that question, perhaps one should ask whether there is a holomorphic equivalence between the ball and the polydisc. If there is such an equivalence, then the other question need not be answered; if there is no equivalence, then perhaps there is no Riemann mapping theorem.

As Poincaré discovered, there is in fact no equivalence between the ball and the polydisc. In the last decade, it has been discovered that, generically, two domains in $\mathbf{C}^2$ are not holomorphically equivalent: that is to say, given a domain U_1, the probability that a domain U_2 which is topologically equivalent to it is also holomorphically equivalent to it is 0. Thus the entire question of the Riemann mapping theorem turns into a rather complex subject in itself—that of classifying domains according to holomorphic equivalence. One of the main things that we shall do in this chapter is to carefully formulate and prove Poincaré's theorem concerning the inequivalence of the ball and the bidisc.

The other big development which occurred near the turn of the century is Hartogs's theorem about analytic continuation. Although we cannot treat this topic in detail here, we should like to discuss it briefly in order to further convey the flavor of several complex variables. So that we may keep matters as simple as possible, we shall avoid formal definitions in this discussion.

A domain U in complex space is called a *domain of holomorphy* if there is a holomorphic function defined on U which cannot be analytically continued to any open set which properly contains U. It turns out (see [KR1]) that the Mittag-Leffler theorem of classical function theory can be used to show that *any* open set in $\mathbf{C}$ is a domain of holomorphy. What Hartogs discovered is that, in dimensions 2 and higher, some domains are domains of holomorphy and others are not. For example, he proved that the domain

$$U = \{(z_1, z_2) : |z_1| < 2 \ , \ |z_2| < 2\} \\ \setminus \{(z_1, z_2) : |z_1| \leq 1, \ |z_2| \leq 1\}$$

has the property that any function holomorphic on U can be analytically continued to

$$\hat{U} = \{(z_1, z_2) : |z_1| < 2, \ |z_2| < 2\}.$$

As previously noted, this theorem has no analogue in one complex variable and begs the question: "Which domains in several complex variables have the property exhibited by Hartogs's phenomenon?" A necessary geometric condition on the boundary of U was discovered by E. E. Levi, and the problem of proving its sufficiency became known as the Levi problem. The Levi problem was solved in considerable generality in the 1950's, but several related questions still remain open.

The point that we have tried to make in this brief essay is that many of the most interesting questions in several complex variables never come up in one complex variable, because they make no sense in that context. What we can hope to accomplish in this chapter is to give the reader a glance at the subject and suggest some of the techniques that can be used effectively. One of our accomplishments will be to give a fairly easy and elegant proof of Poincaré's theorem.

1. Basic Concepts

We introduce here only the few definitions and pieces of notation that we shall need for our immediate purposes. For a more comprehensive introduction to several complex variables, see [KR1, KR2, KR4]. For simplicity of notation, we confine attention to *two* complex variables.

We consider domains $U \subseteq \mathbf{C}^2$, where

$$\mathbf{C}^2 \equiv \mathbf{C} \times \mathbf{C}$$

and (z_1, z_2) is a typical element. Recall that $\mathbf{C}^2$ may be identified with $\mathbf{R}^4$ by

$$\begin{aligned} \mathbf{C}^2 \ni (z_1, z_2) &\longleftrightarrow (x_1 + iy_1, x_2 + iy_2) \\ &\longleftrightarrow (x_1, y_1, x_2, y_2) \in \mathbf{R}^4. \end{aligned}$$

Thus a function of two complex variables is continuous if it is continuous as a function of four real variables, it is continuously differentiable if it is continuously differentiable as a function of four real variables, and so forth.

Let $U \subseteq \mathbf{C}^2$ be a domain, that is a connected open set. Let z and w be complex numbers. We define

$$U_w = \{z_1 \in \mathbf{C} : (z_1, w) \in U\}$$

and

$$U^z = \{z_2 \in \mathbf{C} : (z, z_2) \in U\}.$$

Definition 1.
A continuously differentiable function f on a domain U is said to be *holomorphic* if, whenever $z \in \mathbf{C}$ satisfies $U^z \neq \emptyset$ then $f(z, \cdot)$ is holomorphic on U^z (as a function of *one* complex variable) and whenever $w \in \mathbf{C}$ satisfies $U_w \neq \emptyset$ then $f(\cdot, w)$ is holomorphic on U_w (as a function of *one* complex variable).

Example 1.
The function

$$f(z_1, z_2) = \frac{z_1}{z_2 + 1}$$

is holomorphic on the domain

$$U_1 = \{(z_1, z_2) : z_2 \neq -1\}.$$

The function

$$g(z_1,z_2) = z_1(\bar{z}_2)^2$$

is not holomorphic on any domain. The function

$$h(z_1,z_2) = \Sigma(z_1)^j(z_2)^j$$

is holomorphic on

$$U_2 = \{(z_1,z_2) : |z_1 z_2| < 1\}.$$

□

In several complex variables there are two natural notions of "neighborhood": the ball and the bidisc. If $P = (P_1,P_2) \in \mathbf{C}^2$ and $r > 0$, then we let

$$B(P,r) = \{z \in \mathbf{C}^2 : |z_1 - P_1|^2 + |z_2 - P_2|^2 < r^2\}$$

and

$$D^2(P,r) = \{z \in \mathbf{C}^2 : |z_1 - P_1| < r,\ |z_2 - P_2| < r\}$$

be, respectively, the open ball and open bidisc of center P and radius r. We also let $\bar{B}(P,r)$ and $\bar{D}^2(P,r)$ denote, respectively, the closed ball and bidisc (defined by replacing the symbol $<$ with the symbol $\leq$).

A basic tool in complex analysis is the Cauchy integral formula and its variants. We now derive one such.

□ **Proposition 2.** *If f is holomorphic in a neighborhood $\bar{D}^2(P,r)$ then for all $(z_1,z_2) \in D^2(P,r)$ we have*

$$f(z) = \frac{1}{(2\pi i)^2} \oint_{\partial D(P_1,r)} \oint_{\partial D(P_2,r)} \frac{f(\zeta_1,\zeta_2)}{(\zeta_1 - z_1)(\zeta_2 - z_2)} d\zeta_2 d\zeta_1.$$

PROOF.
The function $f(\cdot, z_2)$ is holomorphic in a neighborhood of $\bar{D}(P_1,r)$. Therefore the one variable Cauchy integral formula yields, for each fixed value of z_2,

$$f(z_1,z_2) = \frac{1}{2\pi i} \oint_{\partial D(P_1,r)} \frac{f(\zeta_1,z_2)}{(\zeta_1 - z_1)} d\zeta_1.$$

Now apply the one variable Cauchy integral formula to the function $f(\zeta_1,z_2)$ in the integrand *in the second variable:* the desired formula follows. □

□ **Corollary 2.1.** *A holomorphic function of two complex variables is infinitely differentiable as a function of four real variables.*

PROOF.
Differentiate under the integral sign, just as in one complex variable. □

□ **Corollary 2.2.** *A function f as in the Proposition has a power series expansion*

$$f(z_1,z_2) = \Sigma a_{jk}(z_1 - P_1)^j (z_2 - P_2)^k$$

which converges absolutely and uniformly on $\bar{D}(P,r)$. The coefficients are given by

$$a_{j,k} = \frac{1}{j!k!} \left(\frac{\partial}{\partial z_1}\right)^j \left(\frac{\partial}{\partial z_2}\right)^k f(P).$$

PROOF.
The proof follows familiar lines from one complex variable. Since f is holomorphic on a neighborhood of $\bar{D}(P,r)$, we may select $r' > r$ so that f is holomorphic on a neighborhood of $\bar{D}(P,r')$. We write

$$f(z) = \frac{1}{(2\pi i)^2} \oint_{\partial D(P_1,r')} \oint_{\partial D(P_2,r')} \frac{f(\zeta_1,\zeta_2)}{(\zeta_1 - z_1)\cdot(\zeta_2 - z_2)} d\zeta_2 d\zeta_1. \tag{$*$}$$

However

$$\begin{aligned}\frac{1}{(\zeta_1 - z_1)} &= \frac{1}{(\zeta_1 - P_1) - (z_1 - P_1)} \\ &= \frac{1}{(\zeta_1 - P_1)} \frac{1}{1 - ((z_1 - P_1)/(\zeta_1 - P_1))} \qquad (**) \\ &= \sum_{j=0}^{\infty} \frac{(z_1 - P_1)^j}{(\zeta_1 - P_1)^{j+1}}.\end{aligned}$$

Similarly,

$$\frac{1}{(\zeta_2 - z_2)} = \sum_{j=0}^{\infty} \frac{(z_2 - P_2)^j}{(\zeta_2 - P_2)^{j+1}}. \tag{$***$}$$

Convergence is absolute and uniform on $\{|z_1 - P_1| \leq r\}$ and $\{|z_2 - P_2| \leq r\}$ respectively. Thus we may substitute $(**)$ and $(***)$ into $(*)$ and switch the order of summation and integration to obtain

$$f(z_1, z_2) = \sum_{j=0}^{\infty}\sum_{k=0}^{\infty}\left[\frac{1}{(2\pi i)^2}\oint_{\partial D(P_1,r)}\oint_{\partial D(P_2,r)} \frac{f(\zeta_1,\zeta_2)d\zeta_2 d\zeta_1}{(\zeta_1 - P_1)^{j+1}(\zeta_2 - P_2)^{k+1}}\right] \times (z_1 - P_1)^j (z_2 - P_2)^k$$

$$\equiv \sum_{j,k=0}^{\infty} a_{j,k} \cdot (z_1 - P_1)^j (z_2 - P_2)^k.$$

Here

$$a_{j,k} = \frac{1}{(2\pi i)^2}\oint_{\partial D(P_1,r)}\oint_{\partial D(P_2,r)} \frac{f(\zeta_1,\zeta_2)d\zeta_2 d\zeta_1}{(\zeta_1 - P_1)^{j+1}(\zeta_2 - P_2)^{k+1}}.$$

The familiar Cauchy theory of one complex variable now tells us that

$$a_{j,k} = \frac{1}{j!k!}\left(\frac{\partial}{\partial z_1}\right)^j \left(\frac{\partial}{\partial z_2}\right)^k f(P).$$

□

For simplicity let us now consider power series

$$S \sim \Sigma a_{j,k}(z_1)^j (z_2)^k$$

expanded about (0,0). Notice that if S converges absolutely at a point (z_1, z_2) then it also converges absolutely and uniformly at the points $(\mu_1 z_1, \mu_2 z_2)$ where $|\mu_1| \leq 1$, $|\mu_2| \leq 1$. This motivates the following definition.

Definition 3.

A domain U is called *complete circular* if whenever $(z_1, z_2) \in U$ and $|\mu_1|$, $|\mu_2| \leq 1$ then $(\mu_1 z_1, \mu_2 z_2) \in U$. Note in passing that a complete circular domain automatically contains 0.

□ **Proposition 4.** *Let*

$$S \sim \Sigma a_{j,k}(z_1)^j(z_2)^k$$

be a power series. Then

$$\mathcal{C} = \{z : S \textit{ converges absolutely in a neighborhood of } z\}$$

is an open, complete circular domain. We call $\mathcal{C}$ the domain of convergence of S.

PROOF.
Combine the definition of $\mathcal{C}$ with that of complete circular domain and the remark preceding it. □

Although we did not assert the converse of the Proposition, it nevertheless suggests that the ball and the bidisc might each be domains of convergence for some power series. Indeed,

$$S \sim \Sigma(z_1)^j(z_2)^k$$

converges precisely on $D^2(0,1)$ and on no larger open set. The power series

$$T \sim \sum_j \left((z_1)^2 + (z_2)^2\right)^j$$

(when rearranged as a sum of *monomials*) converges absolutely on the open ball, and on no larger open set. (Exercise: fill in the details of these assertions.)

One of the themes of this book has been to consider conformal self-maps of domains. It turns out that "conformal" is the wrong notion for functions of several complex variables: if a function of at least three real variables is conformal (in the usual sense of being infinitesimally angle preserving and, up to scaling, length preserving) then that function must be *linear* (exercise: prove this, or see the discussion in [DFN]). The correct morphisms for our purposes are given in the following definition:

Definition 5.

Let U_1 ,U_2 be domains. A function

$$f : U_1 \longrightarrow U_2$$

is called a *biholomorphic mapping* if

$$f = (f_1(z_1, z_2), f_2(z_1, z_2))$$

is a pair of holomorphic functions, f is injective, f is surjective, and f^{-1} is holomorphic.

Remark.

It turns out that the condition that f^{-1} be holomorphic is redundant; however this is hard to prove so, for simplicity, we state the condition explicitly. □

Example 2.

Let $U = D^2(0,1)$. Then

$$f(z_1, z_2) = \left(\frac{z_2 - 1/2}{1 - (1/2)z_2}, i \cdot z_1 \right)$$

is a biholomorphic mapping of U to U. □

As in the material on one variable, we let $\mathrm{Aut}(U)$ denote the collection of biholomorphic self-maps of U.

Example.

Let $U = B(0,1)$. Then

$$f(z_1, z_2) = \left(\frac{z_1 - 1/2}{1 - (1/2)z_1}, \frac{(\sqrt{3}/2)z_2}{1 - (1/2)z_1} \right)$$

is a biholomorphic mapping of the domain U. Of course any unitary rotation is also a biholomorphic mapping of U.

Exercise.

If $U \subseteq \mathbf{C}^2$ is a domain then $\mathrm{Aut}(U)$ forms a group under functional composition. □

2. The Automorphism Groups of the Ball and Bidisc

We shall calculate the automorphism groups of $B(0,1)$ and $D^2(0,1)$. We begin with some preliminary material.

□ **Lemma 1.** *If f_j are holomorphic functions on U and $f_j \to f$ uniformly on compact subsets of U, then any derivative*

$$\left(\frac{\partial}{\partial z_1}\right)^{\ell}\left(\frac{\partial}{\partial z_2}\right)^{k} f_j$$

converges to

$$\left(\frac{\partial}{\partial z_1}\right)^{\ell}\left(\frac{\partial}{\partial z_2}\right)^{k} f,$$

uniformly on compact sets.

PROOF.

The argument is the same as in one complex variable: Fix $P \in U$ and choose $r > 0$ such that $\bar{D}^2(P,r) \subseteq U$. Express f_j on $D^2(P,r)$ as a Cauchy integral of f_j on $\partial D^2(P,r)$. The result is then immediate from differentiation under the integral sign. □

□ **Lemma 2.** *If f is holomorphic on a domain $U \subseteq \mathbf{C}^2$ and if f vanishes on an open subset $W \subseteq U$ then $f \equiv 0$ on U.*

PROOF.
Let

$$C = \{z \in U : f(z) = 0 \text{ in a neighborhood of } z\}.$$

Then C is clearly open and nonempty. On the other hand, if $z \in \partial C \cap U$, then f and all its derivatives vanish at z. Thus the power series expansion for f in a neighborhood of z is identically 0. It follows that C has no boundary points in U. Therefore it must hold that $C = U$. □

□ **Lemma 3.** *Let $\mathcal{F}$ be a family of holomorphic functions on a domain $U \subseteq \mathbf{C}^2$ such that*

$$|f(z)| \leq M < \infty, \qquad \forall f \in \mathcal{F}.$$

Then $\mathcal{F}$ is a normal family.

PROOF.
It is enough to see that $\mathcal{F}$ is normal on any closed polydisc $\bar{D}^2(P,r) \subseteq U$. Let $\bar{D}^2(P,r') \subseteq U$ be a slightly larger polydisc. Differentiate the Cauchy integral formula on $D^2(P,r')$; the standard estimates from one variable then give

$$\left| \frac{\partial f}{\partial z_j}(z) \right| \leq \frac{M}{r' - r}, \qquad \forall z \in \bar{D}^2(P,r), \ f \in \mathcal{F}.$$

It follows that $\mathcal{F}$ is an equicontinuous family on $\bar{D}(P,r)$. Therefore the result follows from the Ascoli-Arzela theorem. □

Definition 4.
If $U_1,\ U_2 \subseteq \mathbf{C}^2$ and

$$f : U_1 \longrightarrow U_2$$

is holomorphic then define $\text{Jac}_{\mathbf{C}} f(z)\ , z \in U_1$, (the *Jacobian matrix* of f at P) to be the matrix

$$\begin{pmatrix} \frac{\partial f_1}{\partial z_1}(z) & \frac{\partial f_1}{\partial z_2}(z) \\ \frac{\partial f_2}{\partial z_1}(z) & \frac{\partial f_2}{\partial z_2}(z) \end{pmatrix}$$

Remark.
Notice that $\text{Jac}_{\mathbf{C}} f(z)$ is distinct from the real Jacobian, $\text{Jac}_{\mathbf{R}} f$, of calculus: the latter would be a 4×4 matrix which arises from treating f as a function from a domain in $\mathbf{R}^4$ to a domain in $\mathbf{R}^4$. □

Next we prove a Schwarz lemma in two complex variables. Although it could be proved in an *ad hoc* manner for the ball and the bidisc, the proof in general is not essentially more difficult; it also allows us to recall a nice argument that we saw earlier in Section 3.2.

□ **Proposition 5.** *Let $U \subseteq \mathbf{C}^2$ be a bounded domain and $P \in U$. Let $f : U \to U$ be holomorphic and assume that $f(P) = P$. If*

$$\text{Jac}_{\mathbf{C}} f(P) = id$$

then

$$f(z) \equiv z.$$

PROOF.
Assume without loss of generality that $P = 0$. Seeking a contradiction, we assume that the conclusion is false.

In a neighborhood of $0 \in U$ we may expand f in a power series. The hypotheses about f imply that this power series has the form

$$f(z_1, z_2) = 0 + z + A_m(z) + \cdots$$

where $A_m(z)$ represents the first nonvanishing monomial after z, A_m having degree $m \geq 2$. Observe that $f = (f_1, f_2)$ is an ordered pair and that each summand on the right-hand side of our equation is an ordered pair. Define

$$\begin{aligned} f^1(z_1, z_2) &= f \\ f^2(z_1, z_2) &= f \circ f \\ f^j(z_1, z_2) &= f^{j-1} \circ f, \qquad \forall j \geq 2. \end{aligned}$$

Since U is bounded, the family $\{f^j\}$ forms a normal family (by Lemma 3). Therefore there is a subsequence f^{j_ℓ} converging to a limit function $\tilde{f}$. By Lemma 1, the mth derivatives of f^{j_ℓ} at P also converge to the corresponding derivatives of $\tilde{f}$ at P.

On the other hand, direct calculation shows that

$$f^2 = z + 2A_m + \cdots$$
$$f^3 = z + 3A_m + \cdots$$
$$\text{etc.}$$

and we see that in fact the mth derivatives of f^j at $P = 0$ blow up. The only way to resolve this contradiction is to conclude that $A_m \equiv 0$ near P. Therefore $A_m \equiv 0$ on U. We conclude that $f(z) \equiv z$, as desired. □

□ **Proposition 6.** *Let U be a bounded complete circular domain. If*

$$f : U \longrightarrow U$$

is a biholomorphic mapping and $f(0) = 0$ then f is linear.

Remark.
This is a generalization to two variables of the fact that a conformal map of the disc to the disc which preserves the origin is in fact a rotation. □

PROOF OF THE PROPOSITION.
Let $\theta \in [0, 2\pi)$ and

$$\rho_\theta(z_1, z_2) = (e^{i\theta} z_1, e^{i\theta} z_2).$$

Consider the map

$$g = \rho_{-\theta} \circ f^{-1} \circ \rho_\theta \circ f.$$

Then

$$\text{Jac}_{\mathbf{C}} g(0) = \rho_{-\theta} \circ (\text{Jac}_{\mathbf{C}} f(0))^{-1} \circ \rho_\theta \circ (\text{Jac}_{\mathbf{C}} f(0)).$$

Here we think of $\rho_{\pm\theta}$ as a linear map given by the matrix

$$\begin{pmatrix} e^{\pm i\theta} & 0 \\ 0 & e^{\pm i\theta} \end{pmatrix},$$

hence $\rho_{\pm\theta}$ is its own Jacobian. Since $\rho_{\pm\theta}$ is diagonal, it commutes with all other 2×2 matrices. So

$$\text{Jac}_{\mathbf{C}} g(0) = id.$$

Now the preceding proposition applies and we may conclude that

$$g(z) \equiv z.$$

This means that

$$f \circ \rho_\theta = \rho_\theta \circ f. \qquad (*)$$

Write f in a convergent power series in a neighborhood of 0:

$$f(z) = \Sigma a_{j,k} (z_1)^j (z_2)^k.$$

Then $(*)$ says that

$$\Sigma a_{j,k} e^{i(j+k)\theta}(z_1)^j(z_2)^k = \Sigma a_{j,k} e^{i\theta}(z_1)^j(z_2)^k$$

for all θ. Equating like monomials implies that

$$a_{j,k} = 0$$

unless $j + k = 1$. It follows that f is linear near 0. By the uniqueness of analytic continuation (Lemma 2), f is linear.

□

□ **Proposition 7.** *Let $\phi \in \mathrm{Aut}(D^2(0,1))$. Then there exist $a_1, a_2 \in D(0,1)$, $\theta_1, \theta_2 \in [0, 2\pi)$, and a permutation σ of the set $\{1,2\}$ such that*

$$\phi(z) = \left(e^{i\theta_1} \cdot \frac{z_{\sigma(1)} - a_1}{1 - \bar{a}_1 z_{\sigma(1)}},\ e^{i\theta_2} \cdot \frac{z_{\sigma(2)} - a_2}{1 - \bar{a}_2 z_{\sigma(2)}} \right).$$

PROOF.
Let $\phi(0) = \alpha = (\alpha_1, \alpha_2)$. Define

$$\psi(z) = \left(\frac{z_1 - \alpha_1}{1 - \bar{\alpha}_1 z_1}, \frac{z_2 - \alpha_2}{1 - \bar{\alpha}_2 z_2} \right).$$

Then $g \equiv \psi \circ \phi \in \mathrm{Aut}(D^2(0,1))$ and $g(0) = 0$. It suffices to show that

$$g(z_1, z_2) = \left(e^{i\theta_1} z_{\sigma(1)},\ e^{i\theta_2} z_{\sigma(2)} \right).$$

We know from Proposition 6 that g is linear. So

$$g(z) = \begin{pmatrix} g_{11} & g_{12} \\ g_{21} & g_{22} \end{pmatrix} \begin{pmatrix} z_1 \\ z_2 \end{pmatrix}$$

for some complex constants g_{ij} of modulus not exceeding one. Define, for $k = 2, 3, \ldots$, points $z^{1,k}$, $z^{2,k} \in D^2(0,1)$ by

$$z^{1,k} = \left((1 - 1/k)\overline{\mathrm{sgn}(g_{11})},\ (1 - 1/k)\overline{\mathrm{sgn}(g_{12})} \right),$$

$$z^{2,k} = \left((1 - 1/k)\overline{\mathrm{sgn}(g_{21})},\ (1 - 1/k)\overline{\mathrm{sgn}(g_{22})} \right).$$

Here, if $w \in \mathbf{C}$, we are letting

$$\mathrm{sgn}\ w \equiv \begin{cases} w/|w|, & w \neq 0, \\ 0, & w = 0. \end{cases}$$

Now

$$g(z^{1,k}) \text{ has first component } (1 - 1/k)(|g_{11}| + |g_{12}|)$$

and

$$g(z^{2,k}) \text{ has second component } (1 - 1/k)(|g_{21}| + |g_{22}|).$$

Letting $k \to +\infty$ gives

$$\begin{aligned} |g_{11}| + |g_{12}| &\le 1 \\ |g_{21}| + |g_{22}| &\le 1. \end{aligned} \qquad (*)$$

On the other hand, apply g to $\alpha^k = (1 - 1/k, 0)$ and $\beta^k = (0, 1 - 1/k)$ and let $k \to +\infty$. We conclude that

$$\Big((1 - 1/k)g_{11},\ (1 - 1/k)g_{21}\Big) \longrightarrow \partial D^2(0,1)$$

and

$$\Big((1 - 1/k)g_{12},\ (1 - 1/k)g_{22}\Big) \longrightarrow \partial D^2(0,1).$$

Therefore

$$\begin{aligned} \max\{|g_{11}|,\ |g_{21}|\} &= 1 \\ \max\{|g_{12}|,\ |g_{22}|\} &= 1. \end{aligned} \qquad (**)$$

The only way that $(*)$ and $(**)$ can both be true is if each column of the matrix (g_{ij}) has one entry of modulus one and one entry 0. Say that $|g_{\eta(k),k}| = 1$ for $k = 1,2$, where η is a permutation of $\{1,2\}$. Letting $\sigma = \eta^{-1}$ gives that

$$|g_{j,\sigma(j)}| = 1, \qquad j = 1,2$$

and

$$g_{j,k} = 0 \qquad \text{if} \quad k \neq \sigma(j).$$

Let $g_{j,\sigma(j)} = e^{i\theta_j}$, $j = 1,2$. Then

$$g(z_1, z_2) = (e^{i\theta_1} \cdot z_{\sigma(1)},\ e^{i\theta_2} \cdot z_{\sigma(2)}),$$

as desired. □

An elegant treatment of the biholomorphic self-maps of the unit ball may be found in [RU2]. We give a more elementary treatment here.

□ **Proposition 8.** *If $a \in \mathbf{C}$, $|a| < 1$, then the map*

$$\phi_a(z_1, z_2) = \left(\frac{z_1 - a}{1 - \bar{a} z_1}, \frac{(1 - |a|^2)^{1/2} z_2}{1 - \bar{a} z_1} \right)$$

is a biholomorphic self-map of $B(0,1)$ to $B(0,1)$.

Remark.

The maps ϕ_a are natural generalizations of the Möbius transformations which we used extensively in the first part of the book. □

PROOF OF THE PROPOSITION.
Now

$$|\phi_a(z)| < 1$$

if and only if

$$\left| \frac{z_1 - a}{1 - \bar{a} z_1} \right|^2 + \left| \frac{(1 - |a|^2)^{1/2} z_2}{1 - \bar{a} z_1} \right|^2 < 1$$

if and only if

$$|z_1 - a|^2 + (1 - |a|^2)|z_2|^2 < |1 - \bar{a}z_1|^2$$

if and only if

$$|z_1|^2 + |a|^2 + (1 - |a|^2)|z_2|^2 < 1 + |a|^2|z_1|^2$$

if and only if

$$(1 - |a|^2)|z_1|^2 + (1 - |a|^2)|z_2|^2 < 1 - |a|^2$$

if and only if

$$z \in B(0,1).$$

□

Exercise.

The inverse of the mapping ϕ_a is ϕ_{-a}. □

Recall that a 2×2 matrix

$$M : \mathbf{C}^2 \longrightarrow \mathbf{C}^2$$

is *unitary* if its inverse equals its conjugate transpose. Geometrically, M is unitary if and only if it takes any orthonormal basis of $\mathbf{C}^2$ (over the base field $\mathbf{C}$) to another.

Proposition 9. *If $g : B(0,1) \to B(0,1)$ is biholomorphic and $g(0) = 0$ then g is a unitary rotation.*

PROOF.
Since $B(0,1)$ is complete circular, the map g must be linear (by Proposition 6). Since g must preserve the boundary of $B(0,1)$ (why?), g takes Euclidean unit vectors to Euclidean unit vectors. But this says that g is unitary. □

Now we can characterize the automorphisms of the unit ball.

□ **Proposition 10.** *If $f : B(0,1) \to B(0,1)$ is a biholomorphic mapping then f is the composition of at most two unitary maps and a map of the form ϕ_a.*

PROOF.
Let $w = f(0)$. There is a unitary rotation α such that

$$\alpha(w) = (|w|,0).$$

The map

$$g = \phi_{|w|} \circ \alpha \circ f$$

is therefore a biholomorphic map of $B(0,1)$ to $B(0,1)$ which takes 0 to 0. By Proposition 9, g must be equal to a unitary operator β. Therefore

$$f = \alpha^{-1} \circ (\phi_{|w|})^{-1} \circ \beta.$$

□

Exercise.

The group $\text{Aut}(D^2(0,1))$ acts transitively on $D^2(0,1)$. The group $\text{Aut}(B(0,1))$ acts transitively on $B(0,1)$. □

The ball and the bidisc are the two most fundamental domains in $\mathbf{C}^2$. Both are contractible, both have transitive automorphism groups, and both are domains of convergence for power series. It came therefore as a great surprise when Poincaré proved that there is no biholomorphic mapping

$$\Phi : D^2(0,1) \longrightarrow B(0,1).$$

He proved this result by first showing that if such a Φ existed then the induced map

$$\text{Aut}(D^2(0,1)) \ni h \longrightarrow \Phi \circ h \circ \Phi^{-1} \in \text{Aut}(B(0,1))$$

would be a group isomorphism. He then analyzed $\text{Aut}(D^2(0,1))$ and $\text{Aut}(B(0,1))$ *as groups* and showed that they could not be isomorphic. Indeed, the connected component of the identity in $\text{Aut}(B(0,1))$ is not commutative while that for $\text{Aut}(D^2(0,1))$ *is* commutative. We have given a sufficiently detailed description of the automorphism groups that the interested reader may carry out Poincaré's program as an exercise.

We shall instead approach the biholomorphic equivalence problem using some of the geometric ideas developed in this monograph. The next section is devoted to that topic.

3. Invariant Metrics and the Inequivalence of the Ball and the Bidisc

We continue to use the symbol D to denote the unit disc in $\mathbf{C}$. Let $U \subseteq \mathbf{C}^2$ be a domain and $P \in U$. Define $(D,U)_P$ to be the holomorphic functions $f : U \to D$ such that $f(P) = 0$. Define $(U,D)_P$ to be the holomorphic functions $f : D \to U$ such that $f(0) = P$. We are now going to define the Carathéodory and Kobayashi metrics. Since we are working in a two-dimensional space, we can no longer specify a metric as a scalar-valued function on the domain. In fact a metric will measure the length of a vector at a point.

Definition 1.

If $P \in U$ and $\xi \in \mathbf{C}^2$ define the *Carathéodory length* of ξ at P to be

$$F_C^U(P,\xi) = \sup\{|\mathrm{Jac}_{\mathbf{C}} f(P)\xi| : f \in (D,U)_P\}.$$

Definition 2.

If $P \in U$ and $\xi \in \mathbf{C}^2$ define the *Kobayashi length* of ξ at P to be

$$F_K^U(P,\xi) = \inf\{|\xi|/|g'(0)| : g \in (U,D)_P,\ g'(0) \text{ is a scalar multiple of } \xi\}.$$

Here and throughout, $|\ |$ denotes Euclidean length.

The way that we define metrics is still motivated by Riemann's philosophy. If $\gamma : [0,1] \to U$ is a continuously differentiable curve, we define its Kobayashi length to be

$$\ell_K(\gamma) = \int_0^1 F_K^U(\gamma(t), \gamma'(t))dt.$$

Notice that we are integrating the lengths, in the metric, of the tangent vectors to the curve. The Carathéodory length of a curve is defined similarly.

One of the basic properties that we shall prove is that holomorphic mappings decrease distance in the Carathéodory and Kobayashi metrics. In several variables we express this assertion as follows.

□ **Proposition 3.** *Let U_1 and U_2 be domains and*

$$f : U_1 \longrightarrow U_2$$

be a holomorphic mapping. If $P \in U_1$ and $\xi \in \mathbf{C}^2$ we define

$$f_*(P)\xi = \text{Jac}_{\mathbf{C}} f(P)\xi.$$

Then

$$F_C^{U_1}(P,\xi) \geq F_C^{U_2}(f(P), f_*(P)\xi)$$

and

$$F_K^{U_1}(P,\xi) \geq F_K^{U_2}(f(P), f_*(P)\xi).$$

PROOF.
We give the proof for the Carathéodory metric. The proof for the Kobayashi metric is similar.

Choose $\phi \in (D, U_2)_{f(P)}$. Then $\phi \circ f \in (D, U_1)_P$. Hence

$$\begin{aligned} F_C^{U_1}(P,\xi) &\geq |(\mathrm{Jac}_{\mathbf{C}}(\phi \circ f)(P))\xi| \\ &= |\mathrm{Jac}_{\mathbf{C}}\phi(f(P)) \circ (\mathrm{Jac}_{\mathbf{C}} f(P))\xi| \\ &= |\mathrm{Jac}_{\mathbf{C}}\phi(f(P))(f_*(P))\xi|. \end{aligned}$$

Taking the supremum over all ϕ gives

$$F_C^{U_1}(P,\xi) \geq F_C^{U_2}(f(P), f_*(P)\xi).$$

□

□ **Corollary 3.1.** *If f is a biholomorphic map then f preserves both the Carathéodory and the Kobayashi metrics; that is, the inequalities in the Proposition become equalities.*

PROOF.
Obvious. □

Exercise.

Check that the Proposition implies that f decreases the lengths of curves. That is, if γ is a continuously differentiable curve in U_1 and $f_*\gamma \equiv f \circ \gamma$ is the corresponding curve in U_2 then show that $\ell_C(f_*\gamma) \leq \ell_C(\gamma)$ and $\ell_K(f_*\gamma) \leq \ell_K(\gamma)$. □

We use Proposition 3 to define two interesting new invariants. These invariants were trivial in one complex variable, but now they provide crucial information.

Definition 4.

Let $U \subseteq \mathbf{C}^2$ be a domain and $P \in U$. The *Carathéodory indicatrix* of U at P is

$$\mathbf{i}_P^C(U) = \{\xi \in \mathbf{C} : F_C^U(P,\xi) < 1\}.$$

The *Kobayashi indicatrix* of U at P is

$$\mathbf{i}_P^K(U) = \{\xi \in \mathbf{C} : F_K^U(P,\xi) < 1\}.$$

In words, the indicatrix is the "unit ball," in the indicated metric, of vectors at P.

□ **Proposition 5.** *Let $f : U_1 \to U_2$ be a biholomorphic mapping of domains in $\mathbf{C}^2$. Say that $f(P) = Q$. Then*

$$\mathrm{Jac}_{\mathbf{C}} f(P) : \mathbf{i}_P^C(U_1) \longrightarrow \mathbf{i}_Q^C(U_2)$$

and

$$\mathrm{Jac}_{\mathbf{C}} f(P) : \mathbf{i}_P^K(U_1) \longrightarrow \mathbf{i}_P^K(U_2)$$

are linear isomorphisms.

PROOF.
Since f is distance decreasing in the Kobayashi metric, $\mathrm{Jac}_{\mathbf{C}} f(P)$ maps $\mathbf{i}_P^K(U_1)$ into $\mathbf{i}_Q^K(U_2)$. But the same observation applies to

$$\mathrm{Jac}_{\mathbf{C}}(f^{-1})(Q) = (\mathrm{Jac}_{\mathbf{C}} f(P))^{-1} ;$$

it maps $\mathbf{i}_Q^K(U_2)$ into $\mathbf{i}_P^K(U_1)$. Thus $\mathrm{Jac}_{\mathbf{C}} f(P)$ is a linear isomorphism of $\mathbf{i}_P^K$ to $\mathbf{i}_Q^K$ as claimed.

The proof for $\mathbf{i}_P^C$ is identical. □

□ **Proposition 6.** *Let $B = B(0,1)$. Then we have*

$$\mathbf{i}_0^K(B) = B.$$

PROOF.
Let $\phi \in (B, D)_0$. If η is any Euclidean unit vector in $\mathbf{C}^2$ then consider the function

$$h(\zeta) \equiv \phi(\zeta) \cdot \eta.$$

Here "·" denotes the usual inner product of 2-vectors. We have that h maps the disc to the disc and $h(0) = 0$. By the Schwarz lemma of one variable,

$$|h'(0)| \leq 1.$$

Since this inequality holds for any choice of η, we conclude that

$$|\phi'(0)| \leq 1.$$

Now if ξ is any vector in $\mathbf{C}^2$ then it follows from the preceding calculation that

$$\begin{aligned} F_K^B(0,\xi) &= \inf\{|\xi|/|\phi'(0)| : \phi \in (B,D)_0\} \\ &\geq |\xi|. \end{aligned}$$

On the other hand, the map

$$\phi_0(\zeta) \equiv \frac{\zeta}{|\xi|}\xi$$

satisfies $\phi_0 \in (B,D)_0$ and $\phi_0'(0)$ is a positive multiple of ξ. Therefore

$$F_K^B(0,\xi) \leq |\xi|/|\phi_0'(0)| = |\xi|.$$

We conclude that

$$F_K^U(0,\xi) = |\xi|,$$

hence that

$$\mathbf{i}_0^K(B) = B.$$

□

□ **Proposition 7.** *Let* $\Omega = D^2(0,1)$. *Then*

$$\mathbf{i}_0^K(\Omega) = \Omega.$$

PROOF.
Define the projections

$$\pi_1(z_1, z_2) = z_1 \qquad \text{and} \qquad \pi_2(z_1, z_2) = z_2.$$

Let $\eta = (\eta_1, \eta_2) \in \mathbf{C}^2$ be any vector. By Proposition 3, we have that

$$F_K^\Omega(0, \eta) \geq F_K^{\pi_1(\Omega)}(\pi_1(0), (\pi_1)_*\eta) = F_K^D(0, \eta_1).$$

But the Schwarz lemma of one variable tells us easily that the last quantity is just $|\eta_1|$. A similar argument shows that

$$F_K^\Omega(0, \eta) \geq |\eta_2|.$$

We conclude from these two inequalities that

$$F_K^\Omega(0, \eta) \geq \max\{|\eta_1|, |\eta_2|\}.$$

Therefore

$$\mathbf{i}_0^K(D^2) \subseteq D^2.$$

For the reverse inclusion, fix η as above and consider the function

$$\phi(\zeta) = \left(\frac{\zeta\eta_1}{\max\{|\eta_1|, |\eta_2|\}}, \frac{\zeta\eta_2}{\max\{|\eta_1|, |\eta_2|\}} \right).$$

Then it is obvious that $\phi \in (\Omega, D)_0$ and that $\phi'(0)$ is a positive multiple of η. Therefore

$$F_K^{\Omega}(0,\eta) \leq \frac{|\eta|}{|\phi'(0)|} = \max\{|\eta_1|,|\eta_2|\}.$$

The opposite inclusion now follows. □

Exercise.

Verify that $\mathbf{i}_0^C(B) = B$ and $\mathbf{i}_0^C(D^2) = D^2$. □

□ **Theorem 8 (Poincaré).** *There is no biholomorphic mapping of the bidisc D^2 to the ball B.*

PROOF.

Suppose that

$$\Phi : D^2 \longrightarrow B$$

is biholomorphic. Let $\Phi^{-1}(0) = \alpha \in D^2$. There is an element $\psi \in \mathrm{Aut}(D^2(0,1))$ such that $\psi(0) = \alpha$. Consider $g \equiv \Phi \circ \psi$. Then

$$g : D^2 \longrightarrow B$$

is biholomorphic and $g(0) = 0$. We will show that g cannot exist.

By Proposition 5, $\mathrm{Jac}_{\mathbf{C}}\, g(0)$ is a linear isomorphism of $\mathbf{i}_0^K(D^2)$ to $\mathbf{i}_0^K(B)$. But Propositions 6 and 7 identify these as D^2 and B respectively. So we have that

$$\mathrm{Jac}_{\mathbf{C}}g(0) : D^2 \longrightarrow B$$

is a linear isomorphism. However this is impossible. For the segment $\mathbf{i} = \{(t + i0, 1) : 0 \leq t \leq 1\}$ lies in ∂D^2. The linear isomorphism would map $\mathbf{i}$ to a nontrivial segment in ∂B. But B is strictly convex (all boundary points are extreme) so its boundary contains no segments. This is the required contradiction. □

Remark.

It is important to appreciate the logic in this proof. The hypothesized map Φ (and therefore g as well) is *not* assumed to extend in any way to ∂D^2. Indeed, given the very different natures of ∂D^2 and ∂B, we would expect Φ to be highly pathological at the boundary. Our geometric machinery allows us to pass to the linear map $\mathrm{Jac}_{\mathbf{C}}g(0)$, which is defined on all of space. Thus we are able to analyze the boundaries of the domains and arrive at a contradiction. □

An important program begun by Poincaré in the early part of the century can be described as follows: If $\Phi : \Omega_1 \to \Omega_2$ is a biholomorphic mapping of smoothly bounded domains in $\mathbf{C}^2$, then suppose that Φ, Φ^{-1} continue smoothly to the boundaries of Ω_1, Ω_2, respectively, in such a way that the extended functions are diffeomorphisms of the closures of the domains. Using just dimension theory (the number of degrees of freedom in Φ versus the number of degrees of freedom in specifying the two real $(2n-1)$-dimensional boundaries), it can be shown that a number of algebraic relations must be satisfied by the derivatives of functions parametrizing $\partial\Omega_1$ and $\partial\Omega_2$. These relations give rise to differential invariants which can play a vital role in the biholomorphic classification of domains.

It is only since 1974 that a substantial beginning has been made in fleshing out Poincaré's program. For in that year, by way of an extraordinarily difficult proof, C. Fefferman [FEF] showed that for a large class of domains Ω_1, Ω_2 it must be that biholomorphic mappings must extend smoothly to the boundary (in one complex variable this is a classical result that goes back to the thesis of Painlevé). At about the same time, Chern and Moser [CHM] and Tanaka [TAN] created systems for calculating the boundary differential invariants.

More recently, S. R. Bell [BEL] and others have come up with simpler proofs of Fefferman's theorem for a much larger class of domains. While it is still believed that a biholomorphic mapping of any two smoothly bounded domains must extend smoothly to their respective boundaries, we are far from being able to prove this assertion.

Epilogue In complex analysis, geometric methods provide both a natural language for analyzing and recasting classical problems and also a rubric for posing new problems. The interaction between the classical and the modern techniques is both rich and rewarding.

Many facets of this symbiosis have yet to be explored. In particular, very little is known about explicitly calculating and estimating the differential invariants described in the present monograph. It is hoped that this book will spark some new interest in these matters.

APPENDIX ON THE STRUCTURE EQUATIONS AND CURVATURE

1. Introduction

Here we give a brief presentation of the connection between the calculus notion of curvature (see [THO]) and the more abstract notion of curvature which leads to the definition of κ in Chapter 2. A more detailed treatment of some of these matters may be found in [ONE].

First, a word about notation. We use the language of differential forms consistently in this appendix. On the one hand, classically trained analysts are often uncomfortable with this language. On the other hand, the best way to learn the language is to use it. And the context of curvature calculations on plane domains may in fact be the simplest nontrivial context in which differential forms can be profitably used. In any event, this appendix would be terribly clumsy if we did not use forms, so the decision is essentially automatic. All necessary background on differential forms may be found in [RU1] or in [ONE].

2. Expressing Curvature Intrinsically

First we recall the concept of curvature for a smooth, two-dimensional surface $M \subseteq$

$\mathbf{R}^3$. All of our calculations are local, so it is convenient to think of M as parametrized by two coordinate functions over a connected open set $U \subseteq \mathbf{R}^2$:

$$U \ni (u,v) \xrightarrow{p} (x_1(u,v), x_2(u,v), x_3(u,v)) \in M.$$

We require that the matrix

$$\begin{pmatrix} \frac{\partial x_1}{\partial u} & \frac{\partial x_2}{\partial u} & \frac{\partial x_3}{\partial u} \\ \frac{\partial x_1}{\partial v} & \frac{\partial x_2}{\partial v} & \frac{\partial x_3}{\partial v} \end{pmatrix}$$

have rank 2 at each point of U. The vectors given by the rows of this matrix span the tangent plane to M at each point. Applying the Gram-Schmidt orthonormalization procedure to these row vectors, and shrinking U, M if necessary, we may create vector fields

$$E_1 : M \longrightarrow \mathbf{R}^3$$

$$E_2 : M \longrightarrow \mathbf{R}^3$$

such that $E_1(x_1, x_2, x_3)$ and $E_2(x_1, x_2, x_3)$ are orthonormal and tangent to M at each $P = (x_1, x_2, x_3) \in M$. Denote by $T_P(M)$ the collection of linear combinations

$$aE_1(x_1, x_2, x_3) + bE_2(x_1, x_2, x_3), \qquad a, b \in \mathbf{R}.$$

We call $T_P(M)$ the *tangent space* to M at P.

Let $E_3(P)$ be the unit normal to M at P given by the cross product $E_1(P) \times E_2(P)$. The functions E_1, E_2, E_3 are smooth *vector fields* on M : they assign to each $P \in M$ a triple of orthonormal vectors.

Let $\delta_1,\delta_2,\delta_3$ denote the standard basis for vectors in $\mathbf{R}^3$:

$$\begin{aligned}\delta_1 &= (1,0,0),\\ \delta_2 &= (0,1,0),\\ \delta_3 &= (0,0,1).\end{aligned}$$

(Many calculus books call these vectors **i**, **j**, and **k**.) Then we may write

$$E_i = \sum_j a_{i,j}(x_1,x_2,x_3)\delta_j, \qquad i = 1,2,3.$$

The matrix

$$\mathcal{A} \equiv \left(a_{i,j}\right)_{i,j=1}^{3},$$

where the $a_{i,j}$ are *functions* of the space variables, is called the *attitude matrix* of the frame E_1,E_2,E_3. Since $\mathcal{A}$ transforms one orthonormal frame to another, $\mathcal{A}$ is an orthogonal matrix. Hence

$$\mathcal{A}^{-1} = {}^t\mathcal{A}.$$

Definition 1.

If $P \in M, v \in T_P(M)$, and f is a smooth function on M, then define

$$D_v f(P) = \frac{d}{dt} f \circ \phi(t) \Big|_{t=0},$$

where ϕ is any smooth curve in M such that $\phi(0) = P$ and $\phi'(0) = v$. One checks that this definition is independent of the choice of ϕ.

Definition 2.

If

$$\alpha : M \longrightarrow \mathbf{R}^3$$

is a vector field on M,

$$\alpha(P) = \alpha_1(P)\delta_1 + \alpha_2(P)\delta_2 + \alpha_3(P)\delta_3,$$

and $v \in T_P(M)$ then define

$$\nabla_v \alpha(P) = (D_v \alpha_1)(P)\delta_1 + (D_v \alpha_2)(P)\delta_2 + (D_v \alpha_3)(P)\delta_3.$$

The operation ∇_v is called *covariant differentiation* of the vector field α.

Definition 3.

If $P \in M$, $v \in T_P(M)$, we define the *shape operator* (or *Weingarten map*) for M at P to be

$$S_P(v) = -\nabla_v E_3(P).$$

□ **Lemma 4.** *We have that $S_P(v) \in T_P(M)$.*

PROOF.
Now $E_3 \cdot E_3 \equiv 1$ hence

$$\begin{aligned} 0 &= D_v(E_3 \cdot E_3)\Big|_P \\ &= (2 \nabla_v E_3) \cdot E_3\Big|_P \\ &= -2S_P(v) \cdot E_3(P). \end{aligned}$$

Hence $S_P(v) \perp E_3(P)$ so $S_P(v) \in T_P(M)$. □

Notice that the shape operator assigns to each $P \in M$ a linear operator S_P on the 2-dimensional tangent space $T_P(M)$.

We can express this linear operator as a matrix with respect to the basis $E_1(P)$, $E_2(P)$. So S assigns to each $P \in M$ a 2×2 matrix $\mathcal{M}_P$.

The linear operator S_P measures the rate of change of E_3 in any tangent direction v. It can be shown that $\mathcal{M}_P$ is diagonalizable. The (real) eigenvectors of $\mathcal{M}_P$ correspond to the *principal curvatures* of M at P (these are the directions of greatest and least curvature) and the corresponding eigenvalues measure the amount of curvature in those directions.

The preceding observations about $\mathcal{M}_P$ motivate the following definition.

Definition 5.

The *Gaussian curvature* $\kappa(P)$ of M at a point $P \in M$ is the determinant of $\mathcal{M}_P$—the product of the two eigenvalues.

Our aim is to express $\kappa(P)$ in terms of the intrinsic geometry of M, without reference to E_3—that is, without reference to the way that M is situated in space.

To this end, we define *covector fields* θ_i which are dual to the vector fields E_i :

$$\theta_i E_j(P) = \delta_{ij}.$$

Then θ_i may be expressed as linear combinations of the standard basis covectors dx_1, dx_2, dx_3. Indeed, if $\mathcal{A} = (a_{i,j})$ is the attitude matrix then

$$\theta_i = \Sigma a_{i,j} dx_j$$

(remember that $\mathcal{A}^{-1} = {}^t\mathcal{A}$). Thus, for each i, θ_i is a differential form; and the standard calculus of differential forms—including exterior differentiation—applies.

From now on, we restrict attention to 1- and 2-forms acting on tangent vectors to M. Thus any 1-form α may be expressed as

$$\alpha = \alpha(E_1)\theta_1 + \alpha(E_2)\theta_2$$

and any 2-form β may be expressed as

$$\beta = \beta(E_1, E_2)\theta_1 \wedge \theta_2.$$

Now we define covector fields $\omega_{i,j}$, $i,j \in \{1,2,3\}$, by the formula

$$\omega_{i,j}(v) = (\nabla_v E_i) \cdot E_j(P).$$

Here "$\cdot$" is the Euclidean dot product. We think of $\omega_{i,j}$ as a differential 1-form. Notice that, since $E_i \cdot E_j \equiv \delta_{i,j}$, we have for $v \in T_P(M)$ that

$$\begin{aligned} 0 &= D_v(E_i \cdot E_j) \\ &= (\nabla_v E_i) \cdot E_j + E_i \cdot (\nabla_v E_j) \\ &= \omega_{i,j}(v) + \omega_{j,i}(v). \end{aligned}$$

Thus

$$\omega_{i,j} = -\omega_{j,i}.$$

In particular,

$$\omega_{i,i} = 0.$$

If $v \in T_P(M)$ then it is easy to check, just using linear algebra, that

$$\nabla_v E_i = \sum_j \omega_{i,j}(v) E_j, \qquad 1 \leq i \leq 3.$$

We call the $\omega_{i,j}$ the *connection forms* for M. We can now express the shape operator in terms of these connection forms.

□ **Proposition 6.** *Let $P \in M$ and $v \in T_P(M)$. Then*

$$S_P(v) = \omega_{1,3}(v) E_1(P) + \omega_{2,3}(v) E_2(P).$$

PROOF.
We have that

$$
\begin{aligned}
S_P(v) &= -\nabla_v E_3 \\
&= -\sum_{j=1}^{3} (\nabla_v E_3 \cdot E_j) E_j \\
&= -\sum_{j=1}^{3} \omega_{3,j}(v) E_j \\
&= \omega_{1,3}(v) E_1 + \omega_{2,3}(v) E_2,
\end{aligned}
$$

since $\omega_{3,3} = 0$. □

Now we can express Gaussian curvature in terms of the $\omega_{i,j}$.

□ **Proposition 7.** *We have that*

$$\omega_{1,3} \wedge \omega_{2,3} = \kappa\, \theta_1 \wedge \theta_2.$$

PROOF.
We need to calculate $\mathcal{M}_P$ in terms of the $\omega_{i,j}$. Proposition 6 gives that

$$S_P(E_1) = \omega_{1,3}(E_1) E_1 + \omega_{2,3}(E_1) E_2$$

and

$$S_P(E_2) = \omega_{1,3}(E_2) E_1 + \omega_{2,3}(E_2) E_2$$

so the matrix of S_P, in terms of the basis E_1, E_2, is

$$\mathcal{M}_P = \begin{pmatrix} \omega_{1,3}(E_1) & \omega_{2,3}(E_1) \\ \omega_{1,3}(E_2) & \omega_{2,3}(E_2) \end{pmatrix}.$$

We know that $\omega_{1,3} \wedge \omega_{2,3}$, being a 2-form, can be written as $\lambda \cdot \theta_1 \wedge \theta_2$. On the other hand,

$$\begin{aligned} \kappa &= \det \mathcal{M}_P \\ &= \omega_{1,3}(E_1)\omega_{2,3}(E_2) - \omega_{1,3}(E_2)\omega_{2,3}(E_1) \\ &= (\omega_{1,3} \wedge \omega_{2,3})(E_1, E_2) \\ &= \lambda. \end{aligned}$$

Therefore

$$\omega_{1,3} \wedge \omega_{2,3} = \lambda\, \theta_1 \wedge \theta_2 = \kappa\, \theta_1 \wedge \theta_2.$$

□

Our goal now is to express κ using only those $\omega_{i,j}$ with $i \neq 3,\ j \neq 3$. For this we require a technical lemma about the attitude matrix.

□ **Lemma 8.** *We have that*

$$\omega_{i,j} = \sum_k a_{j,k} da_{i,k}, \qquad 1 \leq i,\ j \leq 3.$$

PROOF.
If $v \in T_P(M)$ then

$$\omega_{i,j}(v) = \nabla_v E_i \cdot E_j(P).$$

But

$$E_i = \sum_k a_{i,k}\delta_k.$$

Therefore

$$\nabla_v E_i = \sum_k (D_v a_{i,k})\delta_k.$$

Then

$$\begin{aligned}\omega_{i,j}(v) &\equiv \nabla_v E_i \cdot E_j \\ &= \left(\sum_k (D_v a_{i,k})\delta_k\right) \cdot \left(\sum_k a_{j,k}\delta_k\right) \\ &= \sum_k (D_v a_{i,k}) a_{j,k} \\ &= \sum_k da_{i,k}(v) a_{j,k}.\end{aligned}$$

As a result,

$$\omega_{i,j} = \sum_k a_{j,k} da_{i,k}.$$

□

Now we have reached a milestone. We can now derive the important *Cartan structural equations*, which are the key to our intrinsic formulas for curvature.

□ **Theorem 9.** *We have*

$$d\theta_i = \sum_j \omega_{i,j} \wedge \theta_j, \tag{1}$$

$$d\omega_{i,j} = \sum_k \omega_{i,k} \wedge \omega_{k,j}. \tag{2}$$

PROOF.
We have that

$$\theta_i = \sum_j a_{i,j} dx_j$$

hence

$$d\theta_i = \Sigma da_{i,j} \wedge dx_j.$$

Since the attitude matrix $\mathcal{A}$ is orthogonal, we may solve the equations in Lemma 8 for $da_{i,k}$ in terms of $\omega_{i,j}$. Thus

$$da_{i,j} = \Sigma\omega_{i,k} a_{k,j}.$$

Substituting this into our last formula gives

$$d\theta_i = \sum_j \left[\left(\sum_k \omega_{i,k} a_{k,j}\right) \wedge dx_j \right]$$

$$= \sum_k \left[\omega_{i,k} \wedge \sum_j a_{kj} dx_j \right]$$

$$= \sum_k \omega_{i,k} \wedge \theta_k .$$

This is the first structural equation.

For the second equation, notice that the formula

$$\omega_{i,j} = \Sigma da_{i,k} a_{j,k}$$

implies

$$d\omega_{i,j} = -\Sigma da_{i,k} \wedge da_{j,k} .$$

On the other hand,

$$\begin{aligned}
\sum_k \omega_{i,k} \wedge \omega_{k,j} &= \sum_k \left(\sum_\ell da_{i,\ell} a_{k,\ell} \right) \wedge \left(\sum_m da_{k,m} a_{j,m} \right) \\
&= \sum_k \left(\sum_\ell da_{i,\ell} a_{k,\ell} \right) \wedge \left(- \sum_m da_{j,m} a_{k,m} \right) \\
&= - \left(\sum_k a_{k,\ell} a_{k,m} \right) \cdot \left(\sum_{\ell,m} da_{i,\ell} \wedge da_{j,m} \right) \\
&= - \sum_m da_{i,m} \wedge da_{j,m} .
\end{aligned}$$

In the penultimate equality we have used the fact that $\mathcal{A}^{-1} = {}^t\mathcal{A}$. The result now follows. □

The following corollary will prove critical.

□ **Corollary.** *We have*

$$d\omega_{1,2} = -\kappa\, \theta_1 \wedge \theta_2.$$

PROOF.
The second structural equation gives

$$\begin{aligned} d\omega_{1,2} &= \Sigma \omega_{1,k} \wedge \omega_{k,2} \\ &= \omega_{1,1} \wedge \omega_{1,2} + \omega_{1,2} \wedge \omega_{2,2} + \omega_{1,3} \wedge \omega_{3,2}. \end{aligned}$$

Only the third summand doesn't vanish. But we calculated in Proposition 7 that it equals $-\kappa\, \theta_1 \wedge \theta_2$. □

The Corollary has been our main goal in this subsection. It gives an intrinsic way to calculate Gaussian curvature in the classical setting, hence a way to *define* Gaussian curvature in more abstract settings. We now proceed to develop this more abstract point of view.

3. Curvature Calculations on Planar Domains

Let $\Omega \subseteq \mathbf{C}$ be a domain which is equipped with a metric ρ. Assume for simplicity that $\rho(z) > 0$ at all points of Ω. Define

$$E_1 \equiv \frac{(1,0)}{\rho} \qquad \text{and} \qquad E_2 \equiv \frac{(0,1)}{\rho}.$$

Then

$$\theta_1 = \rho \, dx \qquad \text{and} \qquad \theta_2 = \rho \, dy$$

are the dual covector fields. We define $\omega_{i,j}$ according to the first structure equation:

$$\begin{aligned} d\theta_1 &= \omega_{1,2} \wedge \theta_2, \\ d\theta_2 &= \omega_{2,1} \wedge \theta_1. \end{aligned}$$

We define Gaussian curvature according to the Corollary to Theorem 9:

$$d\omega_{1,2} = -\kappa \, \theta_1 \wedge \theta_2.$$

One can check that these definitions are independent of the choice of frame E_1, E_2, but this is irrelevant for our purposes.

We conclude this Appendix by proving that the definition of curvature which we just elicited from the structural equations coincides with the one given in Section 2.1. First, by the way that we've defined θ_1 and θ_2, we have

$$\begin{aligned} d\theta_1 &= d\rho \wedge dx \\ &= (\rho_x dx + \rho_y dy) \wedge dx \\ &= \rho_y dy \wedge dx \\ &= -\frac{\rho_y}{\rho} dx \wedge \rho \, dy \\ &= -\frac{\rho_y}{\rho} dx \wedge \theta_2. \end{aligned}$$

Similarly,

$$\begin{aligned} d\theta_2 &= d\rho \wedge dy \\ &= (\rho_x dx + \rho_y dy) \wedge dy \\ &= \rho_x dx \wedge dy \\ &= -\frac{\rho_x}{\rho} dy \wedge \rho \; dx \\ &= -\frac{\rho_x}{\rho} dy \wedge \theta_1. \end{aligned}$$

Comparison with the first structural equation gives

$$\omega_{1,2} = -\frac{\rho_y}{\rho} dx + \tau \; dy$$

and

$$\omega_{1,2} = -\omega_{2,1} = -\left(-\frac{\rho_x}{\rho} dy \right) + \sigma \; dx,$$

for some unknown functions τ and σ.

The only way these equations can be consistent is if

$$\omega_{1,2} = -\frac{\rho_y}{\rho} dx + \frac{\rho_x}{\rho} dy.$$

Thus

$$\begin{aligned} d\omega_{1,2} &= -\frac{\partial}{\partial y}\left(\frac{\rho_y}{\rho}\right) dy \wedge dx + \frac{\partial}{\partial x}\left(\frac{\rho_x}{\rho}\right) dx \wedge dy \\ &= \left(-\frac{\rho_{yy}}{\rho} + \frac{\rho_y \rho_y}{\rho^2}\right) dy \wedge dx + \left(\frac{\rho_{xx}}{\rho} - \frac{\rho_x \rho_x}{\rho^2}\right) dx \wedge dy \\ &= \frac{1}{\rho^2}\left(\rho\Delta\rho - (\rho_y)^2 - (\rho_x)^2\right) dx \wedge dy \\ &= \frac{1}{\rho^4}\left(\rho\Delta\rho - (\rho_y)^2 - (\rho_x)^2\right) \theta_1 \wedge \theta_2. \end{aligned}$$

The second structural equation now implies that

$$\kappa = -\frac{1}{\rho^4}(\rho\Delta\rho - |\nabla\rho|^2).$$

On the other hand, in Section 2.1 we defined

$$\kappa = -\frac{\Delta \log \rho}{\rho^2}.$$

We have

$$\frac{\partial}{\partial \bar{z}} \log \rho = \frac{1}{\rho}\frac{\partial \rho}{\partial \bar{z}}$$

and

$$\begin{aligned}
\Delta \log \rho &= 4 \frac{\partial^2}{\partial z \partial \bar{z}} \log \rho \\
&= 4\left(-\frac{1}{\rho^2} \frac{\partial \rho}{\partial z} \frac{\partial \rho}{\partial \bar{z}} + \frac{1}{\rho} \frac{\partial^2 \rho}{\partial z \partial \bar{z}} \right) \\
&= -\frac{1}{\rho^2} \cdot | \nabla \rho |^2 + \frac{1}{\rho} \cdot \Delta \rho.
\end{aligned}$$

It follows that, according to the definition in Section 2.1,

$$\kappa = -\frac{1}{\rho^4} (\rho \Delta \rho - | \nabla \rho |^2).$$

Thus, as claimed, the definition of curvature from Chapter 2 is equal to that which arises from the structural equations.

References

[AHL1] L. Ahlfors, An extension of Schwarz's lemma *Trans. Amer. Math. Soc.* 43 (1938), 359–364.

[AHL2] ——, *Complex Analysis,* 3rd. ed. McGraw-Hill, New York, 1979.

[BEL] S. R. Bell, Biholomorphic mappings and the $\bar{\partial}$-problem, *Ann. Math.*, 114 (1981), 103–112.

[BOAS] R. P. Boas, *Invitation to Complex Analysis,* Random House, New York, 1987.

[DFN] B. A. Dubrovin, A. T. Fomenko, and S. P. Novikov, *Modern Geometry—Methods and Applications*, Springer-Verlag, New York and Berlin, 1984.

[EAH] C. Earle and R. Hamilton, A fixed point theorem for holomorphic mappings, *Proc. Symp. Pure Math.*, Vol. XVI, 1968, 61–65.

[FK] H. Farkas and I. Kra, *Riemann Surfaces*, Springer-Verlag, New York and Berlin, 1979.

[FEF] C. Fefferman, The Bergman kernel and biholomorphic mappings, *Invent. Math.,* 26 (1974), 1–65.

[GAR] J. Garnett, *Analytic Capacy and Measure,* Springer Lecture Notes #297 Springer-Verlag, 1972.

[KOB] S. Kobayashi, *Hyperbolic Manifolds and Holomorphic Mappings* Marcel Dekker, New York, 1970.

[KR1] S. Krantz,*Function Theory of Several Complex Variables,* John Wiley and Sons, New York, 1982.

[KR2] ——, What is several complex variables?, *Amer. Math. Monthly,* 94 (1987), 236–256.

[KR3] ——, Functions of one complex variable, *The Encyclopedia of Physical Science and Technology,* Academic Press, New York, 1987.

[KR4] ——, Functions of several complex variables, *The Encyclopedia of Physical Science and Technology,* Academic Press, New York, 1987.

[LS] L. Loomis and S. Sternberg, *Advanced Calculus,* Addison-Wesley, Reading, 1969.

[MIS] D. Minda and G. Schober, Another elementary approach to the theorems of Landau, Montel, Picard and Schottky, *Complex Variables,* 2 (1983), 157–164.

[MUN] J. Munkres, *Elementary Differential Topology,* Princeton University Press, Princeton, 1963.

[ONE] B. O'Neill, *Elementary Differential Geometry,* Academic Press, New York, 1966.

[PRI] I. Privalov, *Randeigenschaften Analytischer Functionen,* Deutsch Verlag der Wissenschaften, Berlin, 1956.

[RU1] W. Rudin, *Principles of Mathematical Analysis,* 3rd ed., McGraw-Hill, New York, 1976.

[RU2] ——, *Function Theory in the Unit Ball of* $\mathbf{C}^n$, Springer-Verlag, Berlin, 1981.

[THO] G. Thomas, and R. Finney, *Calculus and Analytic Geometry,* Addison-Wesley, Reading, 1984.

[WOL] J. Wolf, *Spaces of Constant Curvature,* 4th ed., Publish or Perish Press, Berkeley, 1977.

INDEX